COMPUTER SCIENCE, TECHNOLOGY AND APPLICATIONS

SCIENTIFIC COMPUTING

STUDIES AND APPLICATIONS

COMPUTER SCIENCE, TECHNOLOGY AND APPLICATIONS

Additional books in this series can be found on Nova's website under the Series tab.

Additional e-books in this series can be found on Nova's website under the eBooks tab.

COMPUTER SCIENCE, TECHNOLOGY AND APPLICATIONS

SCIENTIFIC COMPUTING

STUDIES AND APPLICATIONS

CAJ ERLING
EDITOR

nova
science publishers
New York

Library of Congress Cataloging-in-Publication Data

ISBN 978-1-53612-564-1

Published by Nova Science Publishers, Inc. † New York

CONTENTS

Preface		**vii**
Chapter 1	Exploring the Potential for Scientific Computing on the Java Platform of Scala and Groovy with the Scalalab and Groovylab Environments *Stergios Papadimitriou and Lefteris Moussiades*	**1**
Chapter 2	The Structure and Dynamics of Meetup Social Networks *Marija Mitrović Dankulov and Jelena Smiljanić*	**33**
Chapter 3	Efficient Numerical Tools for Solving the Nonlinear Schrödinger Equation *Vladimir Lončar, Ivana Vasić and Antun Balaž*	**63**
Index		**159**

PREFACE

Stergios Papadimitriou and Lefteris Moussiades begin Chapter One by comparing and contrasting some aspects of ScalaLab and GroovyLab, the MATLAB-like environments for the Java Virtual Machine. Additionally, this chapter highlights the strengths and weaknesses of the two environments while examining the dilemma of whether to use dynamic typing or static typing for scientific programming. Next, Chapter Two by Marija Mitrović Dankulov and Jelena Smiljanić express how computational techniques might be used to determine the structure and dynamics of Meetup social groups. Lastly, Vladimir Lončar, Ivana Vasić, and Antun Balaž review parallel numerical algorithms and programs based on the Crank-Nicolson split-step semi-implicit method in order to solve the nonlinear partial differential equations of the Schrödinger type.

As explained in Chapter 1, ScalaLab and GroovyLab are both MATLAB-like environments for the Java Virtual Machine based on Scala and Groovy languages respectively. They present similar user interfaces and functionality to the user. They also share many Java scientific libraries and native code libraries.

From the programmer's point of view though, they have significant differences. Chapter 1 compares some aspects of the two environments and highlights some of the strengths and weaknesses of Scala vs Groovy for scientific computing. The discussion also examines some aspects of the dilemma of using dynamic typing vs static typing for scientific programming. The performance of the Java platform is continuously improved at a fast pace. Today Java can effectively support demanding high-performance computing and scales well on multicore platforms. Thus, both systems can challenge the performance of the traditional C/C++/Fortran scientific code with an easier to use and more productive programming environment.

Computational social science is an emerging interdisciplinary field whose main focus is on investigating human and social dynamics using computational modeling and data analytics. Collective behavior in social systems is of special interest due to its complex nature which demands the use of computational techniques for uncovering and understanding mechanisms that underlie its emergence.

In Chapter 2, the authors demonstrate how computational techniques can be used to quantify the structure and dynamics of Meetup social groups. Dynamics and structure of any Meetup social group strongly depend on the association of its members with the community. Chapter 2 maps the data about the group members and their participation in the group events onto bipartite networks and use tools of complex network theory to analyze their structure and structure of their monopartite projections. The authors explore how the structure of social networks of these groups evolve with time and identify the key topological features that influence collective dynamics and success/survival of the Meetup social group.

As shown in Chapter 3, the nonlinear Schrödinger equation or Gross-Pitaevskii equation plays an important role in several areas of physical sciences. It is used in various forms to describe prominent physical phenomena, such as propagation of light in a nonlinear medium and the dynamics of weakly interacting Bose-Einstein condensates studied in experiments with ultracold quantum gases. Large-scale numerical simulations for realistically large physical systems based on this equation have provided valuable insights into the properties of many interesting phenomena. Furthermore, numerical solutions of this equation are now routinely used for interpretation of experimental data and are indispensable for day-to-day operation of many laboratories, as well as for the design of new experimental setups. Chapter 3, the authors review an efficient numerical algorithm for solving this nonlinear partial differential equation in three spatial coordinates and either real or imaginary time, based on the Crank-Nicolson split-step semi-implicit method. Integration over imaginary time provides a direct access to stationary solutions of the full time-dependent problem. General spatially anisotropic solutions in three spatial variables are considered, as well as dimensionally-reduced problems such as radially or spherically symmetric cases.

Stability and efficiency of the presented method is ensured in the following way. The discretization error in both time and spatial step is of the second order. Moreover, integration over the time variable is split into two steps that are dealt with in a different, but optimized way. In the first step, only local part of the equation is taken into account and integration is performed exactly with respect to it. In this way the nonlinear term is considered in a straightforward, but optimized way that allows for a treatment of strong nonlinearities. In the second step, spatial derivatives are considered separately. Semi implicit nature of the algorithm also contributes to its stability.

This method is numerically implemented using the standard C programming language. Further computational optimization is achieved by employing the state-of-the-art parallelization techniques for different computing platforms: OpenMP (for shared memory multiprocessing), OpenMP/MPI (for distributed memory systems, where each compute node uses OpenMP to maximize performance), CUDA (for single GPGPUs), CUDA/MPI (for distributed memory systems, where each compute node has GPGPU installed), as well as their hybrid combinations. The scalability of the algorithm with the number of processors and compute nodes is discussed.

As applications and extensions of the above developed algorithms, the authors present three relevant examples from the field of ultracold atoms. The first application illustrates how the non-equilibrium dynamics of multicomponent Bose-Einstein condensates can be numerically studied, including the excitation dynamics of a condensate with a coupling of the effective spin and angular momentum in the second application. The third application demonstrates how formation of vortices can be studied in a condensate featuring contact, as well as long-range dipole-dipole interaction.

In: Scientific Computing
Editor: Caj Erling

ISBN: 978-1-53612-564-1

Chapter 1

EXPLORING THE POTENTIAL FOR SCIENTIFIC COMPUTING ON THE JAVA PLATFORM OF SCALA AND GROOVY WITH THE SCALALAB AND GROOVYLAB ENVIRONMENTS

Stergios Papadimitriou* and Lefteris Moussiades
Computer and Informatics Engineering Department,
Eastern Macedonia and Thrace Institute of Technology, Kavala, Greece

Abstract

ScalaLab and GroovyLab are both MATLAB-like environments for the Java Virtual Machine based on Scala and Groovy languages respectively. They present similar user interfaces and functionality to the user. They also share many Java scientific libraries and native code libraries.

From the programmer's point of view though, they have significant differences. This chapter compares some aspects of the two environments and highlights some of the strengths and weaknesses of Scala vs Groovy for scientific computing. The discussion also examines some aspects of the dilemma of using dynamic typing vs static typing for scientific programming. The performance of the Java platform is continuously improved at a fast pace. Today Java can effectively support demanding high-performance computing and scales well on

* Corresponding Author Email: sterg@teikav.edu.gr.

multicore platforms. Thus, both systems can challenge the performance of the traditional C/C++/Fortran scientific code with an easier to use and more productive programming environment.

1. Introduction

At the earlier days of computing scientific software was written in Fortran that was optimized for speed but it was a pure language from the software engineering point of view. Later C and C++ gained in popularity, since without sacrificing speed they offer much richer software development environments. However, they have also major drawbacks as the lack of portability, the lack of standard set of libraries for common applications (e.g., multithreading) and the complexity of pointer arithmetic and multiple inheritance. The Java platform has successfully addressed these drawbacks and fulfilled these gaps and has become of superb popularity both in academic and industrial environments. However, the first versions of the Java Virtual Machine illustrated pure performances for scientific computing, due to the interpretation of the bytecodes. However, the development of the JIT (Just-In-Time) compilation technology and the elaborate design of the Java's HotSpot compiler changed radically that situation. Today, Java directly rivals the performance of C/C++ code and it is a practical alternative for demanding scientific computation.

The recently introduced ScalaLab [14] scientific programming environment for the Java Virtual Machine (JVM) leverages the statically-typed Scala object-oriented/functional language [10]. It provides MATLAB-like syntax that is used to construct scripts that are then compiled by ScalaLab for execution on the JVM.

The GroovyLab environment is based on the Groovy dynamic language for the Java platform [4]. The underlying mechanisms in GroovyLab are very different from ScalaLab, primarily due to the dynamic character of Groovy.

The Scala language supports the implementation of simple, coherent and efficient MATLAB-like interfaces for many Java scientific libraries. These interfaces are compiled within the core of ScalaLab. The Groovy language also provides mechanisms to access easily and elegantly many Java scientific libraries, but these mechanisms are quite different from those provided by ScalaLab.

ScalaLab and GroovyLab are open-source projects and can be obtained from https://github.com/sterglee/scalalabandhttps://github.com/sterglee/ GroovyLab, respectively. Both ScalaLab and GroovyLab can be installed easily. The only prerequisite is the installation of the Java 8 (or newer) runtime (which is free). We supply scripts for launching these systems for all the major platforms. The general high-level architecture of ScalaLab is depicted in Figure 1 and is described in

[14]. The architecture of GroovyLab is shown in Figure 2 and is very similar to that of ScalaLab, except GroovyLab uses the Groovy programming language rather than the Scala programming language. It is also a successor of jLab that is described in [9].

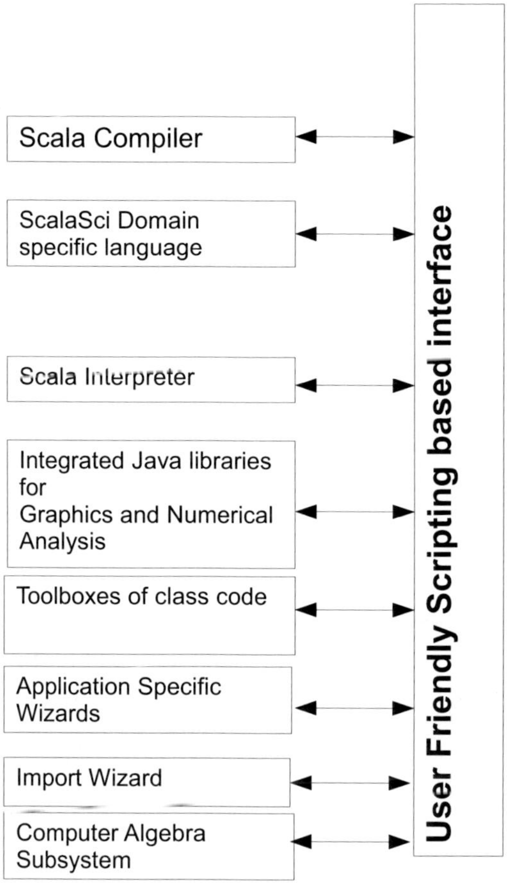

Figure 1. The architecture of ScalaLab.

In this article some important similarities and differences between ScalaLab and GroovyLab will be examined. The user interfaces of the two systems are similar. Also, the exploited Java scientific libraries and native code high performance libraries are about the same. The major differences emerge when code is developed for these systems. They present different scripting languages for writing applications and are also very different when designing and implementing libraries.

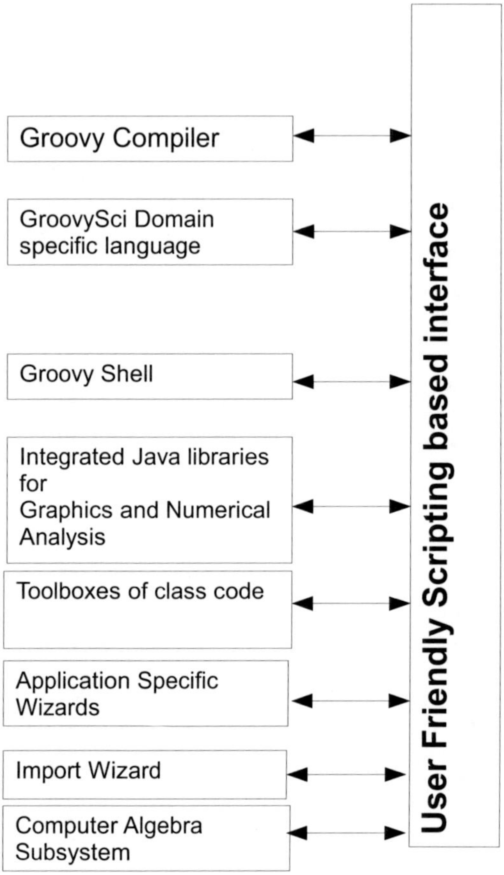

Figure 2. The architecture of GroovyLab.

This chapter proceeds as follows: Initially the frameworks for developing matrix libraries in Scala and in Groovy will be examined (Section 2). Next the implementation of high-level mathematical operators in ScalaLab and GroovyLab are discussed (Sections 3 and 4). The functional programming abilities of the two environments are then briefly examined (Section 5). Both systems provide the user with flexible scripting environments; the main features of these environments are then compared (Section 6). Compile-time metaprogramming is a powerful feature offered both by Groovy and Scala; an example is then presented to demonstrate how compile-time metaprogramming can be used to expand the syntax of GroovyLab without any run-time performance penalties (Section 7). The *C for-loop* of ScalaLab that is implemented using Scala's macros is also discussed at this section. A few aspects of the ScalaLab and GroovyLab environments that are important for scientific computation are then presented and compared. Next performance related issues are discussed and benchmarking results are presented (Section 8). Finally, the paper concludes with remarks concerning the relative strengths and weakness of each system.

2. Matrix Design in Scalalab and GroovyLab

In this section some main features of the Scala and Groovy languages that are used to facilitate the utilization of the Java scientific libraries are described and compared. These features are presented in the context of providing support for a MATLAB-like syntax for matrix manipulation and the utilization of underlying Java libraries to provide the implementation of the matrix functionality.

2.1. Matrices in ScalaLab

The general architecture for interfacing with Java libraries in ScalaLab is illustrated in Figure 3. Below we describe these components.

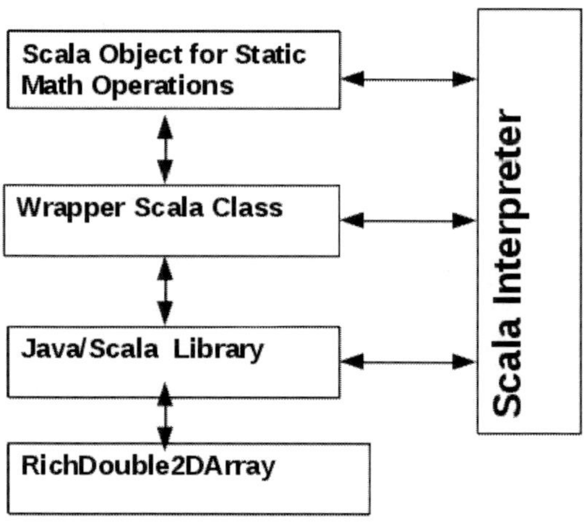

Figure 3. The general architecture of interfacing Java libraries in ScalaLab.

The Java/Scala Library

The *Java library* module in Figure 3 corresponds to the Java code of the library that performs the main numerical calculations. Some examples of Java libraries are the EJML library (https://github.com/lessthan optimal/ejml), an efficient library for the basic linear algebra operations, the Apache Common Maths (http://commons.apache.org/proper/ commons-math/) a rich library that covers plenty of numerical analysis tasks and the MTJ (Matrix toolkits for Java, https://github.com/ fommil/matrix-toolkits-java) that wraps some functionality of the powerful JLAPACK package.

Up-to-now fewer scientific libraries have been developed in Scala compared to Java. Concerning linear algebra and the fundametal numerical analysis tasks, the *Breeze* library (https://github.com/scalanlp/breeze), is currently the more complete one and is competent even compared to established Java libraries, as for example the Apache Common Maths.

It should be noted that the Scala interpreter can also use the native interfaces of each library.

The Wrapper Scala Class (WSC)

The *Wrapper Scala Class (WSC)* aims to provide a simpler interface to the more essential functionality of the Java library, e.g., matrices *A* and *B* can be simply added as *A+B*, rather than invoking the cumbersome *A.plus(B)*. The wrapper Scala class Matrix represents a one-indexed matrix and is based on the NUMAL library [6]. The wrapper class *Mat* is a zero-indexed matrix implemented in Scala. It borrows functionality from the JAMA Java package, a "classic" small package that provides the most basic operations. Some other wrapper classes exist that interface the functionality of important Java libraries as e.g., the *EJML.Mat* class based on the EJML library and the Apache Common Maths library. We implement separate zero-based and one-based indexed *Matrix* classes for two reasons: a) obtaining maximum implementation efficiency and b) the one-indexed *Matrix* class is a wrapper for the NUMAL library routines and it is the only one indexed Matrix class. We feel that is rather inconvenient to mix zero and one indexing styles at the same *Matrix* class. The *scalaSciMatrix* trait factors out the common functionality of the many zero indexed matrix classes.

The Scala Object for Static Math Operations

The *Scala Object for Static Math Operations (SOSMOs)* provides overloaded versions of basic routines for our new Scala matrix types. For example, it provides an overloaded version of the *sin* method that accepts an instance of our *Mat* Scala class as an argument (i.e., *sin(B)* where *B* is a *Mat* instance).

Each SOSMO object implements a large set of mathematical operations. The rationale behind these objects is to facilitate the switching of the Scala interpreter to a different set of libraries. The interpreter simply needs to import the corresponding SOSMO objects in order to switch functionality. In essence, ScalaLab reprograms its set of routines to a different library by importing a library specific wrapper object. GroovyLab could also use a similar mechanism but since Groovy is dynamic and allows run-time metaprogramming it is easier to reprogram the method hooks dynamically.

The top-level mathematical functions for the zero-indexed matrices e.g., *rand0(int n, int m)*, *ones0(int n)* etc., return the matrix representation associated with the currently utilized library. A matrix object can refer to different matrices depending on the library. The "switching" of libraries is performed by creating a different, new Scala interpreter that imports the corresponding libraries with the aid of the specially designed SOSMOs Scala objects. For example, the *StaticMathsJAMA* object, performs important initializations for the JAMA library and the *StaticMathsEJML* utilizes the Efficient Java Matrix Library (EJML). The utilization of the JAMA library is accomplished by creating a Scala interpreter that imports the *StaticMathsJAMA* SOSMO object while for the EJML the *StaticMathsEJML* is imported. The ScalaLab user can easily switch different underlying Java libraries.

The Rich Double 2D Array

The *RichDouble2DArray* is the "super" Matrix class of ScalaLab. It implements mathematical routines that expose the best aspects of various Java scientific computing libraries. It is however independent of any particular utilized library. By convention, utility routines that do not end in 0 or 1 return *RichDouble2DArray* objects. For example, *rand(), zeros(), ones()* etc., all construct *RichDouble2DArray* objects. Furthermore, the extensibility of *RichDouble2DArray* is leveraged with implicit conversions in order to provide its rich functionality to standard two dimensional Java/Scala arrays.

2.2. Matrices in GroovyLab

Some libraries such as the *Apache Common Maths* library, the *JAMA* library and the *NUMAL* library, use a two-dimensional double array matrix representation. The GroovyLab *Matrix* class also uses the same underlying matrix representation, therefore their routines are readily accessible without any conversion. Some other libraries use different matrix representations. For example, JLAPACK uses a one-dimensional double array representation, in which the matrix storage layout is in column based order (i.e., Fortran-like). In these cases, *O(N)* conversion routines (where *N* is the number of matrix elements) are required before using the methods of these libraries. However, mathematical routines with much higher complexity than simple linear complexity (e.g., matrix factorization, Singular Value Decomposition, eigenvalue computations) benefit from such libraries. The EJML library also uses a one-dimensional "flat" matrix representation, in either row or column based order. EJML is one of the more efficient pure Java numerical libraries, one reason being the proper setup for effective caching that the one-dimensional storage representation presents.

b. It provides many useful **static methods** that are usually overloaded to handle many different types. For example, the *sin()* method is overloaded to handle input from a Matrix, a two-dimensional double array and a one-dimensional double array:

Matrix sin(Matrix a)
double [][] sin(double [][] a)
double [] sin(double [] a)

Static importation of all the static methods of the Matrix class is performed by GroovyLab before any code is executed with the GroovyShell (is the component that executes Groovy scripts), therefore we can write *sin(x),* where *x* can take many possible types, e.g., *Matrix, double [][], double [], double.* Therefore, the Matrix class provides much of the same functionality provided by the Scala Objects for Static Math Operations described in the previous section.

2.3. Sparse Matrices

ScalaLab and GroovyLab provide extensive support for sparse matrices using the Java implementation of the techniques presented in [18] (the source is supplied free from the authors). Both systems also implement classes that wrap the sparse routines with a higher level and elegant syntax. Clearly, GroovyLab exploits Groovy's facilities for building Domain Specific Languages (DSLs) and ScalaLab those of Scala. Also, the open source project *Matrix Toolkits for Java (https://github.com/fommil/matrix-toolkits-java)* offers an effective framework for handling sparse matrices with Java classes. These classes are integrated in the build-in libraries of both ScalaLab and GroovyLab and can be utilized with MATLAB like convenience.

3. Designing High-Level Operators in ScalaLab and GroovyLab

Both Scala and Groovy provide the ability to define operators that function on operands of arbitrary types. Groovy's support for operator definition is more limited than Scala's and is restricted to a set of well-known symbolic operators.

3.1. Defining Operators in ScalaLab

ScalaLab's support for operator definition is best demonstrated with an example. The class used as an example for ScalaLab is the *EJML.Mat* class.

The *SimpleMatrix* class of the Efficient Java Matrix Library (EJML, http://code.google.com/p/efficient-java-matrix-library/) implements mathematical operations in an object-oriented way while maintaining the immutability of the operands. For example, to multiply matrix F and x the *mul* method on F can be invoked as in *F.mul(x).* The Java-like method calls are not very elegant though. For example the matrix calculation $F\ P\ F' + Q$ is executed *as F.mult(P).mult(F.transpose()).plus(Q)* instead of the much clearer *F*P*F~ + Q* that is performed in ScalaLab.

The *scalaSci.EJML.Mat* (abbreviated *Mat)* class in ScalaLab wraps the EJML *SimpleMatrix* class and provides the Scala support for high-level MATLAB-like operations.

In Scala, operators on objects are implemented as *method calls,* even for primitive objects like integers (i.e., *int* type). However, although operators are treated syntactically and semantically as method calls, at the code generation phase the Scala compiler treats the usual arithmetic operators on primitive types by generating direct efficient bytecode. Thus, Scala mathematical expressions have speeds similar to those of Java. Operator characters are valid method names. A familiar operator symbol (e.g., '+') can be used to define a method that implements the operator. Infix operators are implemented as methods that have a single parameter that is of the type of the second operand. For example, *a*5* corresponds to *a.*(5). Prefix* operators such as +, -, !, ~ are implemented by prepending the word: *unary_* , to construct the corresponding method name.

P*ostfix* operators are implemented in Scala as methods that take no arguments. For example, implementing a method named ~ on the Matrix class could be used to indicate Matrix transposition. So, the transpose of *A* would be obtained using the ~ operator as *A*~.

Scala provides the *apply* method to provide support for a subscript-like operator for indexed object access that appears similar to both method invocation and array element access. For example for the *EJML.Mat* class, the apply method can be used to provide access the *(row, col)* element of the matrix. The method is implemented, by calling the corresponding routine *get()* of the EJML library, as follows:

```
def apply(row: Int, col: Int) = {
 sm.get(row, col)
 }
```

The method is then invoked as *M(i, j)* to access the element of the matrix in the ith row and jth column of *M*. The *apply* method can also be overloaded in order to obtain a subrange of a matrix as well.

In a similar fashion, Scala provides the *update* method to support indexed object assignment. For the *EJML.Mat* class, the update method can be implemented to assign a value to a specific element of the matrix. The method is implemented as:

```
def update(row: Int, col: Int, value: Double): Unit = {
 sm.set(row, col, value)
 }
```

The *update* method is then invoked as *M(i,j) = 9.8* to assign the number 9.8 to the *i-jth* element of M.

ScalaLab provides an implementation of the colon (:) operator, to support the MATLAB colon operator for creating vectors as the following example:

```
var t = 0::0.02::45.9
```

This *expression returns* a *scalaSci.Vec* type for *t*. To implement such syntax, *implicit conversions* are combined with the *token cells* approach [16, 17].

Methods with names ending with the ':' character are invoked on their right operand, passing in the left operand. For example, the above expression is evaluated as *(45.9.::(0.02)).::(0)*. Since the *Double* class does not have a ::

method, it is implicitly converted to a *MATLABRangeStart* object. The *MATLABRangeStart* object retrieves the receiver's value (i.e., 45.9) with its constructor and stores it as the ending value of the range. The *MATLABRangeStart* class has a method named :: that processes the increment parameter (i.e., 0.02). Finally, the method :: creates a *MATLABRangeNext* object passing itself, i.e., the *MATLABRangeStart* object, as the argument. The :: method of the *MATLABRangeNext* receives as a parameter the starting value of the range (i.e., 0). Therefore, it has all the information (i.e., start, increment, end) to construct and return the vector *t*.

MATLAB-like indexing/assignment is implemented easily by defining overloaded versions of *apply()* that operate on vectors. For example, to evaluate the expression *M(2::4::20, 3::2::100)*, the implicit conversions mechanism converts the arguments to vectors. The apply method is then invoked which will extract from the vector arguments the necessary start, step and increment values.

3.2. Defining Operators in GroovyLab

In contrast to Scala, Groovy provides support for operator overloading only for a specific set of operator symbols. The positive side of this restriction is that we enforced to avoid unusual operator symbols, conforming to the familiar ones (e.g., `*` for multiplication, '<<` for left shift etc.). To support operator overloading, Groovy supplies a standard mapping of operator to the implementing method. For example, the "+" operator corresponds to the *plus* method. From Java only these methods can be used, but from Groovy either the operators or their corresponding methods may be used.

We provide implementations of the Groovy operator methods in pure Java code for both efficiency's sake and to use them from Java as methods. Scala operators cannot be implemented directly in Java (at least without considering the internal details of the Scala compiler) because Java does not support operators as method names. It would also be difficult to call them. In order to call the methods corresponding to Scala operators from Java, the synthetic names that the Scala compiler creates for the operators must be used, which is very inconvenient.

Operator support in Groovy is supplied by implementing the appropriate corresponding method. For example, the *plus* method is implemented to provide the addition operator "+", the *minus method* is implemented to provide the subtraction operator "-" and the *multiply* method is implemented to provide the multiplication "*" operator. This operator-method mapping approach is not as flexible as Scala's "methods as operators" approach that permits the user to define arbitrary symbols as operators. These predefined methods in Groovy can be implemented in Java and (as noted) are implemented in GroovyLab in Java for the sake of efficiency. The syntactic convenience of using the operator rather than the method is only applicable in Groovy though. For example, *x+100* is a valid expression in Groovy, where *x* is a Matrix object, but in Java matrix addition is performed using *x.plus(100)*.

Another example of Groovy's syntactic elegance is demonstrated in the implementation of matrix indexing and assignment operators. The subscript indexing operator in Groovy is overloaded by implementing the *getAt()* method and the subscript assignment operator is overloaded with the *putAt()* method. Groovy has built-in support for integer ranges via the .. operator, so the syntax 2..5 can be used to create an *IntRange* instance. The concept of a step can be implemented by defining a class named *IntRangeWithStep* that inherits from the Groovy range class *IntRange*. This class is used in GroovyLab to write elegant MATLAB like constructs, such as:

```
x = rand(50, 100)
row = 5; col = 2
xr = x[1..4+row, col..4*col] // take a range
xrb = x[(1..20).by(2), (1..30).by(3)] // like, x(1:2:20, 1:3:30)
x[(1..40).by(5), 1..2] = 44.
```

Since the predefined *IntRange* type does not have a *by()* method, the Groovy compiler handles 1..20 as *IntRangeWithStep*. The *by()* method stores the step argument. Finally, elements are accessed using the subscript access operator implemented with the *getAt()* method and the subscript assignment operator that is implemented by the *putAt()* method.

Similarly, the *DoubleRangeWithStep* class that extends *ObjectRange* is provided by *GroovyLab and* implements the *step()* method in order to return a vector. Thus we can write

x = (0.5..50).step(0.01) as equivalent to MATLAB's *0.5:0.01:50.*

After examining both Scala's and Groovy's support for operator overloading, it can be concluded that while GroovyLab's syntax is convenient it cannot provide the MATLAB-like syntax that can be implemented in ScalaLab.

3.3. Matrix Operations Performance

The performance of the basic indexing and assignment operations in GroovyLab's Matrix class when Groovy is statically compiled is similar to that of Scala's two-dimensional matrices. Other Matrix range operations are slower in GroovyLab, even though many of the submatrix operations are performed in Java. Specifically, for operations involving large submatrices the speed is about the same, but performance is about three times better in ScalaLab than in GroovyLab when many small submatrices are processed. In the latter case, the GroovyLab implementation involves much more dynamic Groovy code, rather than the faster Java code, hence the performance penalty.

4. Defining Operators for User Types: Implicit Conversions vs the Metaobject Protocol

Both Scala and Groovy provide mechanisms to handle the situation where an operator is applied to a type in which there isn't a specific operator that matches the argument types. Implicit conversion of argument types at compile time is supplied in Scala to address these potential type mismatches. Groovy calls methods in objects indirectly through their metaclasses. By adjusting trhe metaclasses through run-time metaprogramming, these sorts of type violations can be handled easily.

4.1. Implicit Conversions in Scala

Returning to the matrix *Mat* class example in Scala, when the compiler detects the addition operator '+' on a *Double* object *d* that adds a *Mat* object *M*, i.e., *d* + *M*, it encounters a type error because there is no method defined on the predefined *Double* type that adds a *Mat* instance to a *Double* (and there cannot be one since *Mat* is a user defined type). A similar error occurs when a Mat instance is added to a double array.

Scala provides implicit conversions [10, 11, 12] to address these sorts of type issues. When an operation is not defined for a type, the compiler will try to apply any available implicit conversions in order to transform the type into a type for which the operation is valid.

The concept of implicit conversions is of fundamental importance in the construction of high-level mathematical operators in ScalaLab. Implicit conversion in ScalaLab is used with many classes, e.g., *RichNumber*, *RichDouble1DArray*, and *RichDouble2DArray* classes.

For example, the *RichNumber* class is implemented to support implicit conversions in Scala related to the Double class. The RichNumber class models extended Number capabilities of accepting operations with all the relevant classes of ScalaLab, e.g, with *Mat*, *Matrix*, *EJML.Mat, MTJ.Mat* and generally whatever class we need to process.

Suppose that we have:

var a = 2.0+rand(5,4)

The 2.0 is transformed by the Scala compiler to a *RichNumber* object, that defines an operation to add a Matrix and the addition can be performed by the addition operator implementation.

As *RichNumber* enriches simple numeric types, *RichDouble1DArray*enhances the *Array[Double]*type and *RichDouble2DArray*the *Array[Array[Double]]* type. For example the following code is valid in Scala:

var a = Ones(9, 10) // an Array[Array[Double]] filled with 1s
var b = a+10 // add the value 10 to all the elements returning b as RichDouble2DArray
*var c = b + a*89.7 // similarly using implicit conversions this computation proceeds normally*

In the next section we continue by describing the corresponding implementations of high-level mathematical operators in the context of GroovyLab. Although similar functionality as in ScalaLab can be achieved in GroovyLab, the underlying approaches are very different.

4.2. The Meta-Object Protocol in Groovy

In Groovy as in many other dynamic languages, the implementation of high-level mathematical operators for the standard language types is based on the *Metaobject protocol. The Meta-object protocol* forms the basis of metaprogramming in Groovy that is used to implement dynamic method invocation [4]. This protocol is the means by which dynamic functionality can be added to classes at runtime. Dynamic behavior is added to classes and objects in Groovy using the *MetaClass* machinery.

In dynamic languages, methods can be injected into a class by adding methods to its *MetaClass*. These added methods are then available globally on all objects of the class. In the case of Groovy, metaprogramming can be used to add methods, properties, constructors and static methods to classes at runtime. New methods in Groovy can be added to both Groovy classes and Java classes. Groovy supports metaprogramming at both the class and object level.

For example, the *Number* class of the standard Groovy's library does not implement the addition operator to support adding an array to a Number instance. Metaprogramming can easily be used to define a method on the number class to support this addition operation and adding it to the MetaClass of the Number class, as illustrated in the following Groovy code:

```
// define the operation: Number + double [],
// at the MetaClass of the Number's class
Number.metaClass.plus =
    {
// the double [] m array denotes the input parameter
 double [] m ->

// call a Java routine for the operation

res =
groovySci.math.LinearAlgebra.LinearAlgebra.plus(delegate, m) // calls Java code
    res
}
```

The keyword delegate refers to the current object, i.e., the Number object. Also, since Groovy's bytecode is somewhat slow for numeric calculations (actually when the static compilation feature is not used), GroovyLab intermixes Java code to attain improved performance.

Although this mechanism is seemingly easier to implement than Scala's implicit conversions, it requires indirect method calls through the MetaClass machinery and imposes additional method invocation overhead. However, Groovy's approach is more flexible in that it supports changing which method will be executed at runtime. This flexibility is useful if implementations from different libraries are used interchangeably, say for example when it is desired to use an eigenvalue decomposition from a different library than the one supplied by default.

5. Functional Programming: Functions vs Closures

A core concept in any language that supports functional programming is the provision of functions as 'first-class citizens'. In object-oriented languages like Scala and Groovy functions are objects. They can be passed as arguments to other functions, be a return value from a method and have a concise literal definition syntax. Most programming languages now support first-class functions because they vastly improves the readability and understandability of the source code by allowing behavior to be captured, assigned to variables and passed to functions. For instance, the major new feature of Java 8 is the support of functional programming with *lambda expressions.*

5.1. Functions in Scala

Scala has the concept of both *functions* and *closures.* Functions are like static methods that do not belong to any class/object.

An example of a function definition in Scala is:

*def cube(x: Double) = x*x*x*

Scala also has a literal function syntax that can be used to assign a function to a variable. For example, the function above can be declared as follows:

*val cube = (x:Double) => x*x*x*

The function variable 'cube' can be invoked as if it were the function definition above; that is as in the invocation *cube(3).*

A function in Scala that refers to a variable outside of its scope is called a *closure.* These variables from outside the scope of function are referred to as *free* variables. An example of the definition of a Scala closure is:

val raiseIt = (x: Int) => Math.pow(x, power)

Here 'power' is a free variable because it is defined outside of the scope of the function. At compile time a variable named 'power' must be in the scope of the literal definition of the *raiseIt* function; otherwise a compiler error will result.

Therefore, Scala differentiates between a function and a closure based on whether or not the definition contains free variables.

5.2. Closures in Groovy

Groovy also supports 'global' function definitions (i.e., functions that are not defined within a class) but these functions are simply static methods of classes imported automatically. Therefore, they do not support the functional programming style.

Groovy, however, provides strong support for closures that are *first-class* objects, that can be passed as parameters and assigned to variables. The syntax of closure definition is different from the definition of methods, for example a simple Groovy closure that implements the *cube* function is:

*def cube = { x -> x*x*x }*

The closure is then invoked as a expected as *cube(3)*.

5.3. Global Function Workspace

Scientific programming environments demand a *global namespace of functions*. Scala and Groovy do not have the concept of globally visible methods; every method must be contained in an object or a class. In both environments though, a global function namespace, can be implemented easily with *static imports*. In Scala objects are imported since these objects encapsulate the static imports. In Groovy static imports are performed as they are in Java. Therefore, the automatic import of static methods provides the appearance of the existence of global methods. For example, the *plot* method appears to be available globally since we import it from the object *scalaSci.plot.plot*. Scala also offers the ability to define *apply* methods for the companion objects of classes. If a class implements the apply method, an instance of the class can essentially be 'executed' like a function as the instance name followed by a list of arguments in parentheses. When the apply method is implemented by a class, a method doesn't need to be imported into the 'global' namespace, it is only necessary to import the class itself. Groovy does not offer a similar ability to essentially execute object instances, but the classic *import static* Java facility can be used to import the appropriate static methods.

6. The Scala Interpreter and the Groovy Shell

An essential component of Scala's scripting framework is the Scala interpreter. The corresponding component in the Groovy scripting framework is the

GroovyShell. This section discusses and compares some aspects of these components.

The overall approach of the Scala interpreter is based on initially compiling the requested code. A Java classloader and Java reflection are then used to run the compiled bytecode and access its results.

In general, Scala's approach to scripting and Groovy's approach to scripting are similar. Scala's approach is more complicated than Groovy's approach though. Groovy's scripting approach is based on detecting the undeclared variables of a script. Groovy then declares them implicitly as an Object and maintains their state in a *binding structure* implemented by means of a Java hashtable. The binding scheme of Groovy is restricted to data variables and closures. It is important that although GroovyShell does not keep objects and classes, it keeps *closure* objects. Therefore, the computation workspace consists not only of data variables but also of defined code. This facilitates functional programming within GroovyLab.

In contrast, the scheme implemented in the Scala interpreter extracts the whole visible computation state as the interpreter's context. ScalaLab binds both data and code objects automatically to this context. Although this is more powerful it imposes difficulties in retrieving the context when we create a new Scala interpreter. In that case it is necessary to replay the commands in order to restore the environment. However, the restoration of the user environment in ScalaLab is performed fastly; the user does not notice the delay of the few Scala compiler runs. Restoration of the computation context also is a somewhat rare operation.

A single compiler instance is used in the Scala's interpreter to accumulate all successfully compiled or interpreted Scala code. The interpretation of a line of code is performed by generating a new object that includes the line of code and has public members to export all variables defined by that code.

The results of the interpreted lines are extracted by using a second "result object" which imports the variables exported by the above object and then exports a single member named "result." To accommodate user expressions that read from variables or methods defined in previous statements, "import" statements are used.

It becomes evident that an effective approach for detecting the variables that a piece of code defines is required. The Scala interpreter utilizes the Scala parser to accomplish the non-trivial task of analyzing variable visibility. The Scala parser is also utilized, in order to detect which variables are used by a code snippet. The values of these variables are then requested from previously executed code. It is unnecessary to request variables from the interpreter that do not appear in a new code snippet, since these values will not change. For example, if *prevVar = 5.5,* is not used at our new script, the interpreter does not request the value of prevVar.

Is point we should contrast the state management of the Scala interpreter with the corresponding one of the GroovyShell. GroovyShell automatically handles variable binding and therefore it is much easier to handle the state of variables in GroovyLab.

The Scala interpreter keeps a list of names of identifiers bound with a var or val declaration (*boundVarNames*) and also a list of function identifiers defined with a def declaration (*boundDefNames*). In addition we can retrieve the last text source code of the program that has been interpreted (*lastLineInterpreted*).

We keep a symbol table of the current *ScalaSci* bound variables. This task is used to graphically display the current work context to the user. We have to keep this external table synchronized with the internal variables binding that the interpreter keeps (i.e., *boundVarValNames*). The current value of each variable is retrieved from the interpreter by issuing a simple command with the name of the variable.

If an identifier is being redefined as a function then it is removed from the variable binding. This is necessary since the namespace of variables and functions in Scala is common [10] and thus the identifier is hidden by the function definition.

The corresponding task in the GroovyShell of displaying to the user the variable's workspace is much simpler, in fact it only requires a simple lookup into the hashtable binding.

The GroovyShell lacks the ability to retain imports from a previously executed script. This can be rather inconvenient in many cases, since these same import statements must be manually incorporated before executing code that depend on them. For that reason, GroovyLab implements simple import buffering, that removes the tediousness of incorporating the import statements explicitly each time a script is executed in GroovyLab.

GroovyLab automatically incorporates some basic import statements into the user's scripts that extend the functionality. These imports allow the user to utilize many GroovyLab functions such as figure() and plot(). The user can also use a code buffer to retain code (both classes and script code) for the current working session.

7. Compile-Time Metaprogramming

Compile-time metaprogramming allows intervening in the compilation process and the implementation of customized structures that do not impose any run-time overhead. This is important in scientific scripting, since it opens up possibilities for implementing efficient convenient high-level structures.

Both languages support compile-time metaprogramming, Groovy with *Abstract Syntax Tree (AST) - Transformations* and Scala with *Scala Macros. Abstract Syntax Trees (AST)* keep a tree like representation of a program in which interior nodes represent programming constructs [8]. Scala macros have not been utilized in ScalaLab yet (actually they are used only to provide a C style *for* loop), so the discussion of compile-time metaprogramming will be limited to an exploration of Groovy's compile-time metaprogramming support.

7.1. Compile-Time Metaprogramming in Groovy

Compile-time metaprogramming provides the ability to modify class and instance behavior at run time. Groovy provides compile-time metaprogramming via *AST transformations* and *compilation customizers* to support altering the behavior of classes and objects at compile time.

AST transformations in Groovy come in two flavors: *local AST transformations* and *global AST transformations*. Local AST transformations are defined using annotations and are applied locally to a program element by annotating the element whose AST is to be altered by the AST transformation. Global AST transformations are essentially provided as services that are applied by the compiler to every class that is being compiled. Local AST annotations are more widely used than global AST transformations.

Groovy ships with a number of pre-defined local AST transformations. Examples include the @Immutable AST transformation that generates code to make a class immutable and the @Log AST transformation which injects a logger into a class. Of specific interest to GroovyLab is the *@CompileStatic AST transformation*. This AST transformation is applied to Groovy code that is known not to utilize Groovy's dynamic programming capabilities. When the @CompileStatic AST transformation is applied to a method the compiler checks to see that the source code can indeed be statically compiled and bypasses Groovy's dynamic dispatch for methods that are invoked within the method implementation. This ability to bypass dynamic dispatch can have a marked impact on performance, which is of the utmost importance in scientific computing. We should note that the recent versions of the Groovy compiler, automatically avoid dynamic dispatch in simple numerical loops; thus in such cases Groovy's performance is similar to Java/Scala, even without the static compilation annotation.

GroovyLab also provides the *@CompileJava AST transformation.* This transformation is applied only to methods and essentially compiles the method directly into Java bytecode. The code is compiled as it appears in the Groovy source code, so the source code of the annotated method must be valid Java and must be completely self-contained (i.e., it must essentially be a static method that contains no references to class attributes). The @CompileJava AST transformation provides the ability to execute Java code from within a GroovyLab script, both for efficiency and as a preferred choice.

Groovy also provides the ability to modify the behavior of the compiler using compilation customizers. Compilation customizers are applied by the compiler to the AST corresponding to every class that is being compiled. They differ from global AST transformations in that customizers are easier to define than global AST transformations. GroovyLab uses a compilation customizer to convert BigDecimal literals to primitive double to improve performance.

7.2. Compile-Time Metaprogramming in Scala

Scala provides extensive support for compile time metaprogramming, that however is under intense development and therefore is not considered as stable. The project *ScalaMeta* (https://github.com/scalameta/scalameta) aims to provide an advanced "next-generation" metaprogramming suppport. However, currently the Scala macros framework (http://scalamacros.org/) provides already rather effective support for metaprogramming. We utilized macros in ScalaLab for implementing some new elegant constructs. For example, since Scala's for loop is based on for comprehensions [10] and although it is very flexible, it does not compile effectively the simple numerical loops that are common in scientific computing.

The replacement of for loop with while loops is rather clumsy. For that reason, ScalaLab utilizes the *cfor* loop, implemented with Scala macros. That loop presents the same efficiency as while loops. An example of cfor loop follows:

```
// Ikeda map
close("all")
val R = 1; val C1 = 0.4; val C2 = 0.9; val C3 = 6

tic;
var N=2500000;
```

```
var x = new Array[Double](N)
var y = new Array[Double](N)
x(0)=0.12; y(0) = 0.2;

var km = 0;
var tau=0.0; var sintau=0.0; var costau=0.0

cfor (1) ( _<N, _ + 1) { k =>

 km= k -1
 tau = C1-C3/(1+x(km)*x(km)+y(km)*y(km))
 sintau = sin(tau); costau = cos(tau);
 x(k) = R+C2*(x(km)*costau-y(km)*sintau)
 y(k) = C2*(x(km)*sintau+y(km)*costau)

}

var tm = toc();
scatterPlotsOn(); // display points only, not the connecting lines
plot(x, y);
title("Scala Time for Ikeda map "+tm);
```

8. Performance

Groovy was initially considered a slow language; Groovy 1.5 was about 200 to 1000 times slower than Java in performing number calculations. The implementers of the language have attained impressive performance improvements in recent versions of Groovy though. A clever implementation of call-site caching in Groovy 1.6, has reduced the performance gap between Groovy and Java to Groovy being about 20 – 50 times slower than Java. Special handling of primitive operations as direct bytecode rather than via dynamic dispatch using the MetaObject protocol has provided significant performance gains, primarily because these optimizations can be applied at compile time. In many cases, Groovy with compiled computational loops and primitive operations runs as fast as Java code (since both emit direct bytecodes). Statically-compiled Groovy, a mode of the Groovy compiler introduced with Groovy 2.0, produces statically compiled code that is as fast as Java, but requires sacrificing the dynamic features of the Groovy language in the source code that is to be statically compiled. Therefore, this mode is most suitable for computationally intensive methods. Groovy 2.0 also exploits the JDK7 *invoke-dynamic* bytecode and the related framework that supports the compilation of dynamic code. However in the current release of Groovy 2.8 the invoke dynamic implementation improved a lot related to Groovy 2.0. Generally code that is compiled with Groovy's invoke dynamic support currently runs somewhat slower than the code that optimizes primitive operations. The speed difference is constantly decreased on each new Groovy release. As both the invoke dynamic support in Groovy improves and Java Virtual Machine implementations better support the invoke dynamic instruction, the static compilation and primitive optimizations features of Groovy become of less importance.

GroovyLab is designed with the goal in mind of performing fast numeric calculations. For example, all of the main mathematical operations in GroovyLab are implemented in pure Java for the sake of efficiency. The important *Matrix* class is also implemented in Java, but it implements the *GroovyObjectSupport* interface, in order to allow flexible overloaded operator syntax, e.g., to use *A+B* instead of something like the cumbersome method *A.plus(B)* to add matrices. The Matrix class also supports many of operations that make use of the native BLAS. This is accomplished with the JBLAS library (http://mikiobraun.github.io/jblas/).

Another performance pitfall with Groovy is the preference of the Groovy compiler to perform default mathematical calculations with *BigDecimal*s objects. Groovy is a flexible, extensible language: that allows GroovyLab to bypass that constraint by performing a *compile-time metaprogramming* transformation. These transformations are a powerful tool to modify the AST representation of the source code that is generated during compilation, before bytecode is generated from it for execution. Since the AST modifications are performed by the compiler during compilation, there is are no run-time performance penalty that result from applying the AST modifications.

In case of ScalaLab, the Scala language is statically-typed and therefore Scala code is compiled to bytecode that runs as fast as Java, sometimes a bit faster sometimes a bit slower, depending on the situation. However, for the advanced features of Scala such as pattern matching, trait inheritance and type parameters, it is difficult to optimize their compilation. The Scala language developers concentrate on these issues and improve the performance of the Scala compiler with each new version of the language.

The following table compares characteristics of *ScalaLab, MATLAB, SciLab* and *GroovyLab.* SciLab (www.scilab.org) is an open source system, similar to MATLAB.

	ScalaLab	MATLAB	SciLab	GroovyLab
Speed	Very Fast, execution speed depends on the Java Runtime, generally faster than MATLAB at script code, but slower for routines implemented as built-in with MATLAB	Very Fast, especially the built-in routines are highly optimized, overall ScalaLab and MATLAB run at comparable speeds and which one outperforms depends on the case	Much slower than ScalaLab (or MATLAB), about 20 to 100 times slower. Newer versions of SciLab however, improved a lot, speed differences are now about 3 to 10 times.	Slower than ScalaLab, about 2 to 5 times slower. However, with statically typed blocks of code, performance is at about the same level as Java/Scala.
Portability	Very portable, anywhere exists installed Java 7 JRE	There exist versions for each main platform, e.g., Windows, Linux, MacOS	There exist versions for each main platform, e.g., Windows, Linux, MacOS	Very portable, anywhere exists installed Java 7 JRE
Open-source	Yes	No	Yes	Yes
User-friendness	Very user friendly	Very user friendly	Very user friendly	Very user friendly
Libraries/Toolbox availability	All the JVM libraries	A lot of toolboxes are available, but generally not free	There exist toolboxes for basic applications but for specialized ones is difficult to find	All the JVM libraries
Documentation	Little yet, and limited to on-line help, since even main code components are in the development process.	Extensive documentation	Sufficient documentation	On-line documentation only

	ScalaLab	MATLAB	SciLab	GroovyLab
Flexibility of the language (i,e, syntax mallability)	The Scala language is designed to be flexible and with very mallable syntax	The syntax of MATLAB is not designed to be extensiible.	SciLab is not designed to be extensible.	The Groovy language is dynamic and different tricks from the Scala's case can form customizable syntax
Development of large applications	Scala has a lot of novel features that can facilitate the development of large applications. ScalaLab applications can run standalone, as any Java code.	The notion of MATLABPATH integrates many MATLAB scripts, something not very scalable	Similar to MATLAB, the SciLab scripts are not well suited for complex applications, but rather they fit well for rapid testing of scientific algorithms	Groovy has a full compiler that can be used to produce standalone code of a large application project
Active user development community	ScalaLab is a new project, and thus up-to-now lacks a large user base	MATLAB has a huge user base	SciLab has a large user base, however much smaller than MATLAB's	GroovyLab is a new project, and thus up-to-now lacks a large user base

9. Benchmarking

In order to access the performance of the GroovyLab and ScalaLab platforms, a variety of mathematical computation algorithms will be examined. These will include matrix computations, Fast Fourier Transforms (FFT), eigen decomposition of a matrix and singular value decomposition of a matrix.

9.1. Matrix Computation Benchmarking

In order to access the efficiency of matrix processing in GroovyLab and ScalaLab, variations of the following MATLAB script will be used.

Array access benchmark in MATLAB

```
N=2000; M = 2000
tic
a = rand(N, M);
sm = 0.0;
for r =1:N,
 sm = 0.0;
 for c=1: M,
 a(r, c) = 1.0/(r+c+1);
 sm = sm+ a(r,c)-7.8*a(r,c);
 end
 end
 tm = toc
```

For the ScalaLab version of this script, ScalaLab clearly outperforms both MATLAB and SciLab. GroovyLab has similar speed when the static compilation is used. With the implementation of *optimized primitive operations* (i.e., later versions of Groovy produce fast code for arithmetic operations since they avoid the overhead of the meta-object protocol) and with the later *invoke dynamic* implementation, Groovy generally is slightly slower than Scala. The reason for the superiority of ScalaLab in terms of scripting speed, is clearly the statically typed design of the Scala language that permits the emission of efficient Java bytecode.

9.2. Fast Fourier Transform Benchmark

The Fast Fourier Transform (FFT) benchmark is performed in ScalaLab using implementations of the FFT from various libraries.

Of these libraries, the Oregon DSP library provides the best performance. Close in performance to this library is the JTransforms (https://sites.google.com/site/piotrwendykier/software/jtransforms) library. Since JTransforms is multithreaded, it will accordingly perform more efficiently with more robust machines (e.g., having 8 or 32 cores, instead of only 4). The tutorial FFT implementation of the classic Numerical Recipes book [7] (with the C/C++ code translated to Java) was also observed to achieve reasonable performance in ScalaLab. Interestingly, we observed that the Oregon DSP and JTransforms FFT routines are nearly as fast as the optimized built-in FFT of MATLAB. We should note that the reported differences in benchmarks are stable, e.g., the relative differences are about the same on different computers, and individual runs show small variations at the results. Contributing to the small variability is that we perform explicitly garbage collection before any benchmark run.

Table 1. Results of some basic benchmarks

	ScalaLab (secs)	**SciLab 5.21 (secs), SciLab 5.5**	**MATLAB 7.1 (secs) MATLAB 2012b**	**GroovyLab (secs)**
Matrix Multiplication with matrix sizes: (2000, 2500) X (2500, 3000)	0.9 secs using Native BLAS combined with Java multithreading	61.8, 5.05	13.05, 0.6	The same with ScalaLab
LU				
1000	0.3	3.13, 2.42	0.36, 0.03	The same as ScalaLab
1500	1.2	3.82, 2.1	1.18, 0.04	As ScalaLab.
2000	2.9	6.42, 1.6	2.72, 0.09	As ScalaLab.
inv				
1000	2.7	12.97, 1.6	1.3, 0.05	As ScalaLab.
1500	7.8	13.14, 2.5	4.5, 0.15	As ScalaLab.
2000	9.31	19.07, 3.2	5.9, 0.3	As ScalaLab.
QR				
1000	1.03	4.3, 4.2	1.2, 0.04	As ScalaLab.
1500	3.7	9.96, 9.9	4.26, 0.2	As ScalaLab.
2000	9.25	19.69, 19.3	9.89, 0.3	As ScalaLab.
Matrix access scripting benchmark	0.03	32.16, 32.67	10.58, 0.32	0,031 static compilation, 0,156 with primitive ops, 0,211 with invoke dynamic
FFT 100 ffts of 16384 sized signal	Oregon DSP: real case: 0.05, complex case: 0.095 Jtransforms: real case: 0.07 complex case: 0.11, Apache Common Maths: complex case: 0.5 Numerical Recipes (Java Translation): real case: 0.09 complex case: 0.12	Real case: 2.32 Complex case: 4.2	Real case: 0.05 Complex case: 0.08	The Java libraries for FFT are the same as ScalaLab's

9.3. Other Benchmarks

Other types of problems such as the eigen decomposition, singular value decomposition, solution of overdetermined systems etc. were examined for the purposes of obtaining GroovyLab and ScalaLab benchmarks. The general conclusion is that ScalaLab is faster than SciLab 5.21 by about 3 to 5 times, but is slower than MATLAB 7.1 by about 2 to 3 times. It is also evident that the routines of JLAPACK for special matrix categories run orders of magnitude faster than routines for general matrices, e.g., for a 1500 by 1500 band matrix with 2 bands above and 3 bands below the main diagonal, the JLAPACK's SVD routines runs about 250 times faster than for a general 1500 by 1500 matrix.

Table 1 summarizes some basic performance results. We should note that often ScalaLab and GroovyLab perform equally well since they call the same Java library routines.

Recent MATLAB versions have improved impressively the performance of Matrix multiplication and of many important routines, as e.g., the SVD computation. However, both ScalaLab and GroovyLab offer the potential to issue commands to the MATLAB engine using the Java/MATLAB interface. Similarly, SciLab scripts can be executed using the Java/SciLab interface. The wiki pages of the ScalaLab and GroovyLab projects describe details and provide examples of these interfacings.

Table 2. SVD performance: Java vs native C code

Matrix Size	Optimized C (gcc, similar is for cl)	ScalaLab/ GroovyLab	Unoptimized C (gcc, similar is for cl)
200X200	0.08	0.15	0.34
200X300	0.17	0.2	0.61
300X300	0.34	0.58	1.23
500X600	3.75	5.06	8.13
900X1000	35.4	51.3	53.3

9.4. Native Code Optimizations

In order to test the JVM performance vs native code performance, an implementation of SVD is used. Both the Microsoft's *cl* compiler of Visual Studio on Windows 8 64-bit and the *gcc* compiler running on Linux 64-bit were used. ScalaLab is based on the Java runtime version: 1.7.0_25 and Scala 2.11 M7, and GroovyLab on Groovy 2.2.1. Again both ScalaLab and GroovyLab perform similarly, since they are based on the same Java code. It has been observed that ScalaLab and GroovyLab perform better than unoptimized C and are even close to optimized C code when performing matrix calculations. Table 2 shows some results.

Conclusion

This article compares some aspects of ScalaLab and GroovyLab, which are both environments for scientific computing that run within the Java Virtual Machine framework. It was demonstrated that both environments can effectively utilize existing Java scientific software. Both can elegantly integrate well-known Java numerical analysis libraries for basic tasks. These libraries are wrapped by either Scala objects in ScalaLab or Groovy objects in GroovyLab and their basic operations are provided to the user with a uniform MATLAB-like interface.

An extension of Scala with MATLAB-like constructs called ScalaSci is the language of ScalaLab and the corresponding language of GroovyLab is GroovySci. Both languages are effective and convenient for both writing small scripts and for developing large production- level applications.

The design of the user interface of ScalaLab and GroovyLab is similar. Both emphasize user friendliness and provide integrated development environment (IDE)-like features such as on-line help, code completion, graphical control of the class-path and a specialized text editor with code coloring facilities that greatly facilitate the development of scientific software.

Future work will concentrate on improving the interfaces to Java basic libraries and on incorporating smoothly other interested libraries (e.g., the parallel COLT library for basic linear algebra, the JCUDA library for supporting the CUDA massively parallel computing framework on NVIDIA graphics cards). Both ScalaLab and GroovyLab explore the *symja* Java Computer Algebra system (*https://code.google.com/p/symja/*). This system implements a wide range of Computer Algebra facilities. Further work is in progress for making work with Computer Algebra easier. This work will include providing better on-line help and code-completion for these routines. These components are of the utmost importance in incorporating this rather complicated libraries into ScalaLab and GroovyLab.

References

[1] Cay Horstmann, Gary Cornell, *Core Java 2*, Vol I Fundamentals, Vol II - Advanced Techniques. Sun Microsystems Press, 8th edition, 2008.

[2] S. Papadimitriou, *Scientific programming with Java classes supported with a scripting interpreter*, IET Software, 1, (2), pp. 48 - 56, 2007.

[3] Papadimitriou S, Terzidis K., *jLab: Integrating a scripting interpreter with Java technology for flexible and efficient scientific computation*, Computer Languages, Systems & Structures (2008), *Elsevier, vol. 35, 2009, pp. 217–240.*

[4] Dierk Konig, Andrew Glover, Paul King, Guillaume Laforge, Jon Skeet, *Groovy In Action*, Manning Publications, 2007.

[5] Simon Haykin, *"Neural Networks and Learning Machines," Third Edition*, Pearson Education, *2009.*

[6] Hang T. Lau, *A Numerical Library in Java for Scientists and Engineers*, Chapman & Hall/CRC, 2003.

[7] William H. Press, Saul A. Teukolsky, William T. Vetterling, Brian P. Flannery, *Numerical Recipes in C++, The Art of Scientific Computing, Second Edition*, Cambridge University Press, 2002.

[8] Alfred Aho, Monica S. Lam, Ravi Sethi, Jefrey D. Ullman, *Compilers, Principles, Techniques, & Tools, Second Edition*, Addison-Wesley, 2007.

[9] Stergios Papadimitriou, Konstantinos Terzidis, Seferina Mavroudi, Spiridon Likothanasis, "Scientific Scripting for the Java Platform with jLab," *IEEE Computing in Science and Engineering (CISE)*, July/August 2009, Vol. 11, No 4, pp. 50-60.

[10] Martin Odersky, Lex Spoon, Bill Venners, *Programming in Scala, Artima eds, third edition*, 2016.

[11] Dean Wampler & Alex Payne, *Programming Scala,* O'Reily, 2009.

[12] Venkat Subramaniam, *Programming Scala – Tackle Multicore Complexity on the Java Virtual Machine,* Pragmatic Bookself 2009.

[13] Thomas Wurthinger, Christaian Wimmer, Hanspeter Mossenblock, *"Array Bounds Check Elimiation for the Java Hotspot Client Compiler,"* PPPJ 2007, September 5-7, 2007, Lijboa Portugal, ACM.

[14] Stergios Papadimitriou, Konstantinos Terzidis, Seferina Mavroudi, Spiridon Likothanasis, "*ScalaLab*: an effective scientific programming environment for the Java Platform based on the *Scala* object-functional language," IEEE Computing in Science and Engineering (CISE), Vol. 13, No 5, 2011, p. 43-55.

[15] E. Anderson, Z. Bai, C. Birschof, S. Blackford, J. Demmel, J. Dongarra, J.Du Croz, A. Greenbaum, S. Hammarling, A. Mckenney, D. Sorensen, *LAPACK Users' Guide*, SIAM, Third Edition, 1999.

[16] Gilles Dubochet, "On Embedding domain-specific languages with user-friendly syntax," In *Proceedings of the 1st Workshop on Domain Specific Program Development*, pages 19-22, 2006.

[17] Gilles Dubochet, *"Embedded Domain-Specific Languages using Libraries and Dynamic Metaprogramming,"* PhD Thesis, EPFL, Suise, 2011.

[18] Timothy A. Davis, *"Direct Methods for Sparse Linear Systems,"* SIAM publishing, 2006.

In: Scientific Computing
Editor: Caj Erling
ISBN: 978-1-53612-564-1

Chapter 2

THE STRUCTURE AND DYNAMICS OF MEETUP SOCIAL NETWORKS

Marija Mitrović Dankulov*[*] and Jelena Smiljanić
Scientific Computing Laboratory,
Center for the Study of Complex Systems,
Institute of Physics Belgrade,
University of Belgrade, Belgrade, Serbia

Abstract

Computational social science is an emerging interdisciplinary field whose main focus is on investigating human and social dynamics using computational modeling and data analytics. Collective behavior in social systems is of special interest due to its complex nature which demands the use of computational techniques for uncovering and understanding mechanisms that underlie its emergence. Here we demonstrate how computational techniques can be used to quantify the structure and dynamics of Meetup social groups. Dynamics and structure of any Meetup social group strongly depend on the association of its members with the community. We map the data about the group members and their participation in the group events onto bipartite networks and use tools of complex network theory to analyze their structure and structure of their monopartite projections. We explore how the structure of social networks of these groups evolve with time and identify the key topological features that influence collective dynamics and success/survival of the Meetup social group.

[*]Corresponding Author Email: mitrovic@ipb.ac.rs.

PACS: 05.65.+b, 89.75.Fb, 89.75.Hc

Keywords: complex networks, social groups, self-organization

1. Introduction

The abundance of social data is the main driving force behind the expansion of a new interdisciplinary fields commonly known as *computational social science* [1,2]. Researchers from different fileds of science including physics, computer science, mathematics, economics, and sociology, make the best use of data, computational techniques and resources to provide us with quantitative description and understanding of complex social systems. Computational methods from statistical mechanics and complex network theory have proven to be a useful tool for quantitative description and uncovering the mechanisms that underlie the collective behavior in social systems [3,4]. These techniques have been udsed to explain the emergence of various collective social phenomena, including segregation [5], cooperation [6], collective emotions [7–10], popularity [11,12], etc.

One manifestation of collective social behavior is a social group, informal community of individuals that emerges through repeated interactions among them. Although evolution and organization of social groups have been subject of interest in sociology and related scientific disciplines for more than a century [13–16], the development of information communication technologies and new quantitative techniques have provided new insight into the structure and dynamics of various social groups. The availability of data enabled the study of human behavior and the emergence of social groups in online environment on a scale that was not possible just two decades ago [8, 17–22]. However, the dynamics and structure of event-driven social groups are still insufficiently explored, mainly due to the lack of data. These groups have a vital place in society since they are a significant part of human social life [23–25]. The dynamics of these groups is very localized in time and space. Furthermore, their members meet and build social interactions at group events, which makes them very different from other offline and online social groups. Recent research on several types of these groups [26,27], has shown that universal features characterize their dynamics and that structure of social relations among their members strongly influences member's activity in the group and thus her association with the community.

In this chapter, we demonstrate how computational techniques from statistical physics and complex network theory can be used to explore in details the structure and dynamics of event-driven social groups. As a case study, we use four different Meetup groups, communities of individuals formed online that get engaged into joint offline activities. Meetup members are organized into thematic social groups, which differ in physical location, type of activity and size. Thus, the analysis of these four groups allows us to detect and describe their universal structural and dynamical features. In particular, we show how the data about members activity in a Meetup group can be mapped onto a bipartite graph, and how this network can be used to obtain the group's social relations. The topological structure of a bipartite network and its monopartite projection reveals important details about social dynamics in these groups. We show that four considered groups have very similar structural properties, which further highlights the fact that universal principles govern dynamics of these groups. These universal principles can be uncovered through quantitative analysis of time series related to individual and collective activity. Our analysis shows that correlations between the activity of individual members are the principal cause of specific structural and dynamical patterns, which are characteristic for Meetup, and presumably other event-driven, social groups.

In Section 2 we describe in detail the four considered Meetup social groups and the data used for empirical analysis. The detailed description of structural properties of bipartite and monopartite weighted social networks is given in Section 3, while the analysis of dynamics of these groups is given in Section 4.

2. Data

Meetup [28] *is an online social networking site that facilitates offline group meetings in various locations around the world.* It enables its members to find and join social groups by their interests and get involved into activities of these groups. Therefore, Meetup groups bring together people with common interests who are usually not acquainted outside of the group's social circle. This fact allows us to study the evolution of group's structure and dynamics from its beginnings, as well as their mutual influence, using computational techniques from complex network theory and statistical physics.

Depending on the interest of its members, a group can be assigned to one of 33 categories, which include careers, hobbies, socializing, health, politics, books, etc. Besides topic and location, these groups also differ in size, event

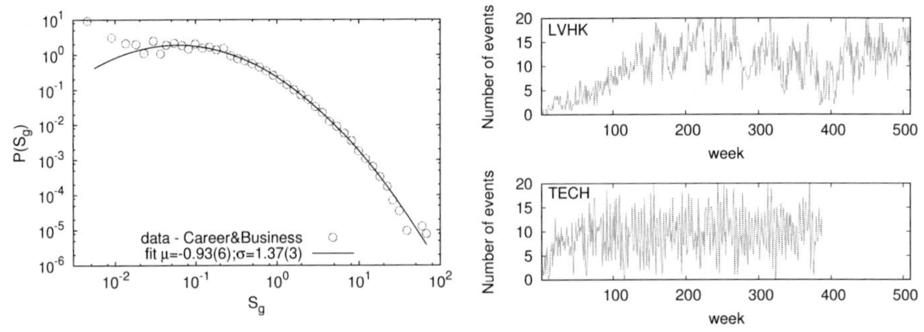

Figure 1. (left) Probability density distribution of group sizes S_g belonging to Meetup category *Career&Business*, which exhibits log-normal behaviour $f(S_g) = \frac{1}{S_g\sigma_g\sqrt{2\pi}}\exp(-\frac{(S_g-\mu_g)^2}{2\sigma_g^2})$. Here S_g is group size rescaled with average group size of category *Career&Business*. (right) Time series of number of events organised per week for two out of four groups we considered in this chapter.

dynamics, and organizational structure. Figure 1(left) shows the probability density distribution of sizes of groups belonging to category *Career&Business*. This distribution exhibits a log normal behavior, which suggests that some multiplicative mechanisms underlie the growth of Meetup social groups. Time series of the number of organized events per week for two different groups, shown in Fig. 1(right), clearly demonstrate the difference in event dynamics: the number of events organized on a weekly basis in two different groups fluctuates differently. Despite their differences, Meetup groups have one thing in common, the way their members prepare and organize their events. Before each event, all members of the group receive an invitation to participate in an event to which they can reply with yes or no. These recordings of event attendance can be then used to analyze members event participation patterns [27] and structure of social network formed between the members of the group.

In this work, we will demonstrate the power of computational techniques by analyzing the data from four different Meetup groups, where each one of them has more than 3000 members and organized events, see Table 1. To obtain more general conclusions, we have selected groups from different categories and with a different type of activities. Specifically, our first group *geamclt* (GEAM) brings together individuals, *foodie thrill-seekers*, who are mostly interested in trying new exotic foods and drinks. Thus, the activities of this group include collective visits to restaurants and bars. The second group (LVHK) consists of hikers who seek excitement and like to socialize through physical activities, while members of our third (PGHF) group like to attend free, or almost free, social events. Our fourth group (TECH) gathers technology professionals interested in networking, entrepreneurship, environmental sustainability, and professional development. We used the Meetup public API to collect the data. For all four groups we compiled the list of all events organized by the group since its beginning, and for each event, we collected the list of members who attended that event. Every member of the Meetup community has a unique identifiers which allows us to follow her activity through time. The detailed description of the data is given in Table 1. To protect the privacy of Meetup users we have conducted a full anonymisation of the data. The anonymised data used for the analysis presented in this work can be found at [29].

Table 1. Description of four Meetup groups studied in this chapter. N_M is total number of group members, while N_E is total number of organized events

Meetup group	Acronym	Category	N_M	N_E
geamclt	GEAM	Food & Drink	5377	3986
pittsburgh-free	PGHF	Socializing	4995	4617
techlifecolumbus	TECH	Tech	3217	3162
VegasHikers	LVHK	Outdoors & Adventure	6061	5096

3. Structure of Meetup Social Groups

The data from event-driven social groups are typically mapped onto monopartite weighted graph [30–32]. In this framework, each member is represented by a network node, while the link weight between two individuals is either raw [33] or normalized [30, 31] number of common events that they have attended together. The problem with this approach is that repeated co-occurrence of two group members at group's events does not necessarily imply social relationship between them. For instance, according to this approach two very active individuals, where by very active we mean persons who attended a significant number of group's events, will have a high probability to meet on one or even several different events. In the monopartite network model, this manifests as a social relationship between these two individuals, although their relationship can be just a consequence of chance.

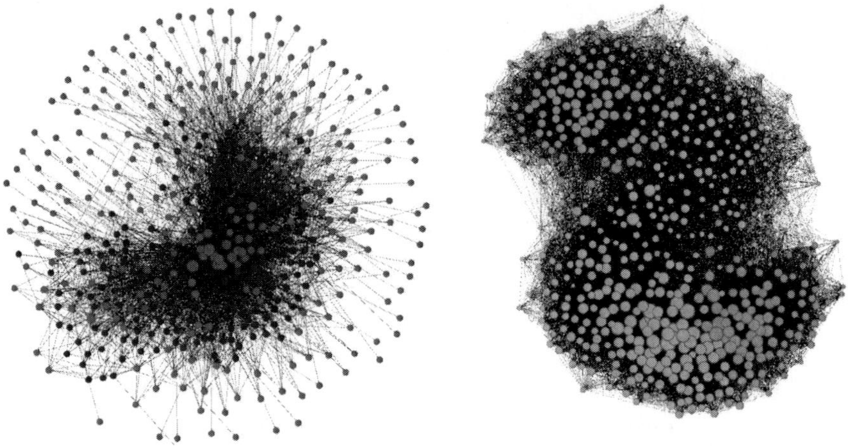

Figure 2. (left) A part of Meetup bipartite network of GEAM community: light gray circles represent nodes from Members partition while nodes from Events partition are represented with white circles. (right) Weighted network of significant social relations between the 10% most active members of LVHK group. Nodes of the same shade of gray belong to the same community.

In our previous work [27], we have shown that relevant social relations in event-driven social group can be obtained by mapping the data onto bipartite network and then use this network to filter out redundant links. First we map the data onto a network with two partitions *Members* (M) and *Events* (E). In bipartite networks, by definition, only links between nodes belonging to two different partitions are allowed. Here, the link between nodes i^M and j^E denotes presence of member i^M at event j^E. This way we obtain binary bipartite networks of size $N_M + N_E$, which can be represented as a block diagonal matrix B, where $B_{i^M j^E} = 1$ if there is a link between member i^M and event j^E, otherwise $B_{i^M j^E} = 0$. Example of bipartite networks of GEAM Meetup group is shown in Fig. 2(left). Bipartite networks created in this way contain all the necessary information about event-driven activity in Meetup groups. For instance, degree of a node q_i^M in Members partition, defined as

$$q^M_{i^M} = \sum_{j^E=1}^{N_E} B_{i^M j^E} , \tag{1}$$

is equal to the number of events that member represented by that node has attended, while the degree of a node belonging to Events partition is equal to the size of corresponding event,

$$q^E_{j^E} = \sum_{i^M=1}^{N_M} B_{i^M j^E} . \tag{2}$$

In order to filter out redundant connections we use technique based on the configuration model of bipartite random graphs (CMBR) [34, 35]. In this filtering technique, CMBR is used for estimation of probability that two members have occurred together on w events by chance, i.e. link π-value. If this chance is low, below some threshold θ, we can claim at confidence level $1-\theta$ that two members have correlated participation in group events, i.e. social relation between them is significant (real) and its strength is w. Before calculating this probability we first need to determine the probability $p_{i^M j^E}$ that member i^M has attended some event j^E by chance, i.e. a probability that there is a link between member i^M and event j^E in bipartite random network. For this we use graphs generated with CMBR and maximal-entropy approach for calculating the link probability in ensemble of networks [36–38]. CMBR graph has the same degree distribution for both partitions as the real network but all other properties, for instance degree-degree correlations, are randomized. Specifically, if we use $\mathcal{G}$ to denote the ensemble of bipartite random graphs with given degree distributions, then probability to have a specific realization of graph G from this ensemble is given by

$$P(G) = \frac{1}{Z} e^{-\sum_{i^M j^E} \alpha_{i^M} q^M_{i^M} + \beta_{j^E} q^E_{j^E}} , \tag{3}$$

where α_{i^M} and β_{j_E} are Lagrangian multipliers and Z is partition function. The partition function for the ensemble of CMBR graphs can be easily calculated (see [27, 35]) and it is equal to

$$Z = \sum_G e^{-\sum_{i^M j^E} (\alpha_{i^M} + \beta_{j^E}) B_{i^M j^E}} = \prod_{i^M j^M} (1 + e^{-(\alpha_{i^M} + \beta_{j^E})}) . \tag{4}$$

The values of Lagrangian multipliers can be calculated using the following equations

$$q^M_{i^M} = -\frac{\partial Z}{\partial \alpha_{i^M}} = \sum_{j^E=1}^{N_E} \frac{e^{-\alpha_{i^M}-\beta_{j^E}}}{1+e^{-\alpha_{i^M}-\beta_{j^E}}} , \tag{5}$$

and

$$q^E_{j^E} = -\frac{\partial Z}{\partial \beta_{j^E}} = \sum_{i^M=1}^{N_M} \frac{e^{-\alpha_{i^M}-\beta_{j^E}}}{1+e^{-\alpha_{i^M}-\beta_{j^E}}} . \tag{6}$$

If we define a coupling parameter $\lambda_{i^M j^E} = \alpha_{i^M} + \beta_{j^E}$ then the link probability between member i^M and event j^E is equal to

$$p_{i^M j^E} = \frac{\partial Z}{\partial \lambda_{i^M j^E}} = \frac{e^{\lambda_{i^M j^E}}}{1+e^{-\lambda_{i^M j^E}}} = \frac{e^{-\alpha_{i^M}+\beta_{j^E}}}{1+e^{-\alpha_{i^M}+\beta_{j^E}}} . \tag{7}$$

The links in CMBR graph are uncorrelated, thus the probability that two members i^M_1 and i^M_2 have attended the same event j^E is simply a product of $p_{i^M_1 j^E}$ and $p_{i^M_2 j^E}$. The probability that two members have co-occurred in w events is given by Poisson binomial distribution

$$P_{i^M_1 i^M_2}(w) = \sum_{Ew} \prod_{j^E \in Ew} (p_{i^M_1 j^E} p_{i^M_2 j^E}) \prod_{j^E \in Ew} (1 - p_{i^M_1 j^E} p_{i^M_2 j^E}) , \tag{8}$$

where Ew is the subset of w events that can be chosen from given E events [36]. Finally, π-value for a link between members i^M_1 and i^M_2 is a probability that these two members have attended together at least $w_{i^M_1 i^M_1}$ events

$$\pi(w_{i^M_1 i^M_2}) = \sum_{w \geq w_{i^M_1 i^M_2}} P_{i^M_1 i^M_2}(w) . \tag{9}$$

If this π-value for a pair of members is high we can assume that their activity in group events is not correlated and that $w_{i^M_1 i^M_2}$ co-occurrences are a result of chance. On the other hand, if the π-value for some pair of members is low, i.e. smaller than θ, we can assume at the level of confidence $1-\theta$ that activity of these members is correlated to a certain extent. This means that the link of weight w between these members can not be created using CBMR graphs. The π-values of weighted links between members of one Meetup group can be then used to filter out the redundant links and create weighted network of significant social relations.

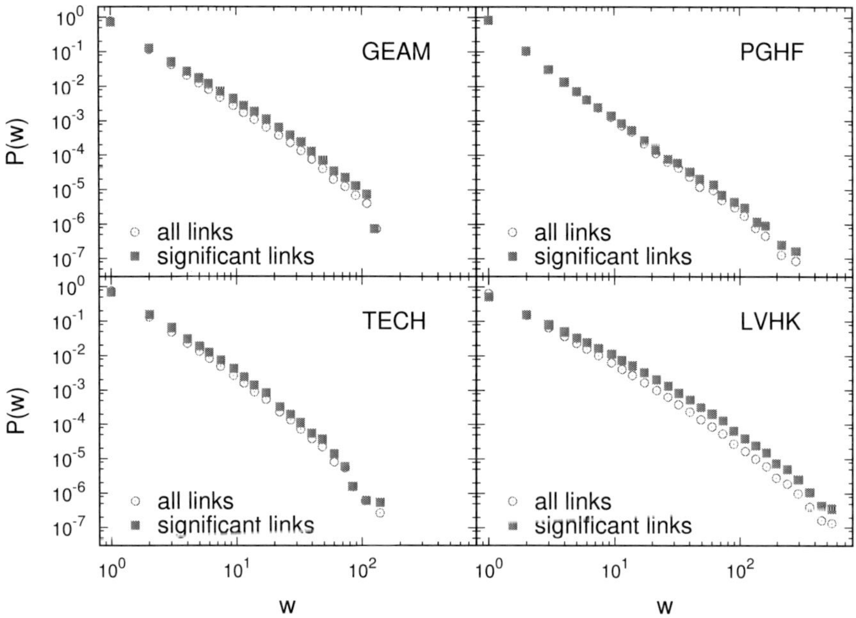

Figure 3. Probability distribution of link weights in a weighted network before and after filtering, for four considered Meetup groups. [Figure right reprinted with permissions from [27]]

We map the data from one Meetup group onto bipartite networks and then apply the described filtering technique to obtain the weighted network of significant social relations between its members at the 95% confidence level ($\theta = 0.05$). This weighted network can be represented with weighted matrix W, where $W_{i_1^M i_2^M}$ denotes the strength of the social relation between members i_1^M and i_2^M. The probability distribution of the number of common events, given in Fig. 3, has power-law behavior for all four networks. Fat-tail distribution of commons is typically found in techno-social systems characterized by high heterogeneity of members activity [7, 17, 27, 33]. The weighted network obtained from a projection of bipartite network onto one of the partitions is usually very dense, i.e. has a huge number of connections per node. We decrease the network density using the described filtering technique. We see from Fig. 3 that although this method puts more emphasis on the links with higher weight, the overall distribution of link weights is preserved.

In our previous work [27], we have examined in detail how the topological properties of member's ego network evolve with the number of attended events. So far, the structure of bipartite and weighted monopartite networks corresponding to Meetup groups has not been studied and described in detail. We quantify the structure of these networks by calculating several local, mesoscopic and global topological measures. The standard topological measure, used for quantifying the connectivity patterns in the networks, is the node degree and its distribution [39]. The degree of a node in Members and Events partitions is defined with Eqs. 1 and 2, while in the weighted monopartite network the node degree is equal to the number of its different links, i.e.

$$q^W_{i_1^M} = \sum_{i_2^M=1}^{N_M} \mathcal{H}(W_{i_1^M i_2^M}) \;, \tag{10}$$

where $\mathcal{H}(W_{i_1^M i_2^M})$ is Heaviside function ($\mathcal{H}(x) = 1$ if $x > 0$ and 0 otherwise). The degree distribution $P(q)$ is a probability that randomly chosen node in a network, bipartite or monopartite, has a value of degree q. The connectivity in bipartite networks is quantified with two-degree distributions, each corresponding to one partition. Besides the degree and its distribution, one more local property is used for the description of local connectivity in a weighted network, node strength. Strength of a node is equal to a sum of weights of all links adjacent to that node, i.e.

$$s^W_{i_1^M} = \sum_{i_2^M=1}^{N_M} W_{i_1^M i_2^M} \;. \tag{11}$$

Degree-degree correlations is another local topological measure with which we capture the mixing patterns in a network [39]. Specifically, with this measure, we quantify node's preference to attach to similar nodes in a network. It can be estimated by calculating the average nearest neighbor degree of a node $\langle q^\kappa_{nn} \rangle$ and examining its dependence on node degree. The average nearest neighbor degree of node with degree q^κ is given by

$$\langle q^\kappa_n n \rangle = \frac{\sum_{j^{\kappa'}} q^{\kappa'}_{j^{\kappa'}}}{q^\kappa} \;. \tag{12}$$

Here κ and κ' denote two different partitions in bipartite networks, in the case of weighted network $\kappa = \kappa' = W$, and $q^{\kappa'}_{j^{\kappa'}}$ is degree of the nearest neighbor node. In uncorrelated networks, the average nearest neighbor degree does not depend on node's degree. If $\langle q_{nn}^{\kappa}\rangle$ is increasing function of q we have assortative mixing, i.e. nodes in this network have preference toward nodes with similar or higher degree, while decreasing function of $\langle q_{nn}^{\kappa}\rangle(q)$ indicates disassortative mixing, large degree nodes are linked to ones with small degree. Social networks are usually characterized by assortative mixing patterns, while biological, technological and techno-social networks typically have disassortative mixing [39]. For weighted networks, one can also define weighted average nearest neighbor degree $\langle q_{nn}\rangle^{W}$, which measures the effective affinity of a node to connect with high- or low-degree neighbors concerning the connection strength. Weighted average nearest neighbor degree of node i^{M} is given by

$$\langle q_{nn}\rangle_{i_1^M}^{W} = \frac{1}{s_{i_1^M}^{W}} \sum_{i_2^M} W_{i_1^M i_2^M} q_{i_2^M}^{M} . \tag{13}$$

Another important measure of topological structure of monopartite networks is clustering coefficient. It measures a probability that two first neighbors of a node are also neighbors. For weighted monopartite social networks we can define non-weighted clustering coefficient of node i^{M}

$$c_{i_1^M} = \frac{1}{q_{i_1^M}^{W}(1-q_{i_1^M}^{W})} \sum_{i_2^M i_3^M} \mathcal{H}(W_{i_1^M i_2^M})\mathcal{H}(W_{i_1^M i_3^M})\mathcal{H}(W_{i_2^M i_3^M}) , \tag{14}$$

and weighted clustering coefficient

$$c_{i_1^M}^{W} = \frac{1}{s_{i_1^M}^{W}(1-q_{i_1^M}^{W})} \sum_{i_2^M i_3^M} \frac{W_{i_1^M i_2^M} + W_{i_1^M i_3^M}}{2} \mathcal{H}(W_{i_1^M i_2^M})\mathcal{H}(W_{i_1^M i_3^M})\mathcal{H}(W_{i_2^M i_3^M}) . \tag{15}$$

While the non-weighted clustering coefficient measures the number of closed triples formed by node and its neighbours, the $c_{i^M}^{W}$ measures local cohesiveness of the node's neighbourhood by taking into account the intensity of interactions between local triplets [40]. Non-weighted $\langle C\rangle$ (weighted $\langle C^{W}\rangle$) clustering coefficient of the whole network is calculated as average value of non-weighted(weighted) clustering coefficients of all nodes in the network. One can also calculate the average clustering coefficient of nodes with degree q^{W} and thus obtain how c and c^{W} depend on q^{W}. This dependences reveal a lot about the hierarchical organisation in network [39, 40], while the averaged values of clustering coefficients provide global information between weights and topology in the network [40]. For instance, the comparison between $\langle C\rangle$ and $\langle C^{W}\rangle$ can reveal much about the importance of clustering in network evolution [40]:

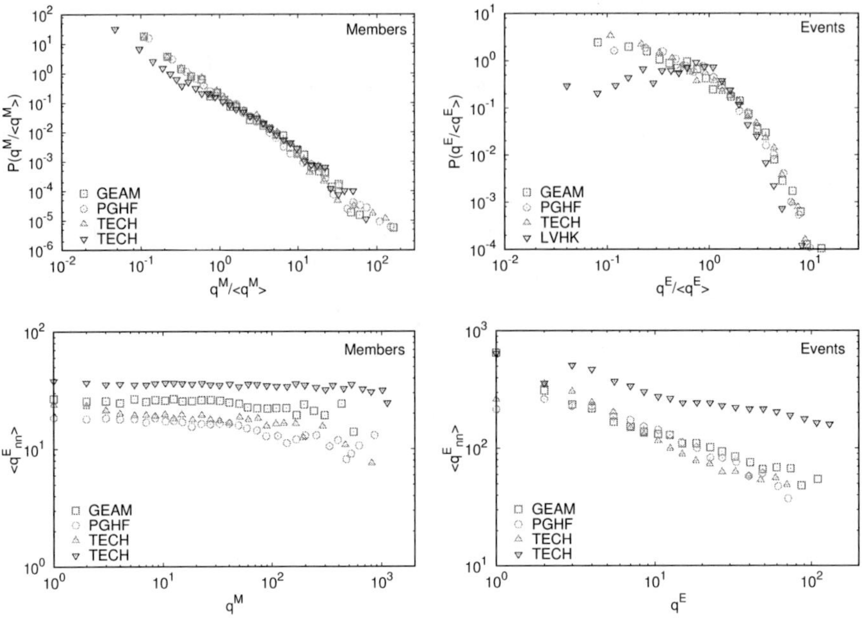

Figure 4. Probability density distribution of rescaled degree for nodes in Members (top left) and Events (top right) partition. The degree is rescaled by the value of an average degree in a given partition. The dependence of average nearest neighbor degree on degree in Members (bottom left) and Events (bottom right) partitions.

$\langle C^W \rangle > \langle C \rangle$ suggests that the processes leading to high value of clustering coefficient have an important role in network evolution.

We first examine the topological structure of bipartite networks corresponding to four considered Meetup groups by calculating the degree distribution for Members, Fig. 4(top left), and Events partition Fig. 4(top right). Degree distribution of Members partition is equivalent to the distribution of the number of attended events studied in [27]. Distributions for all four groups exhibit the same, truncated power-law, behavior

$$P(q^M) \propto (q^M)^{-\gamma} exp\left(-\frac{q^M}{q_0^M}\right), \tag{16}$$

with very similar values of exponent γ ($1.3 \leq \gamma \leq 1.61$) [27]. The observed power-law dependence in Members partition is a result of universal event-driven social dynamics [26, 27]. In event-driven dynamics probability for a member to attend the next event depends exclusively on balance between the number of previous participations and non-participations. Our analysis of the evolution of members ego networks in Meetup social groups has shown that through event attendance members widen and strengthen their social connections and that active members repeat attendance in the group events with a small subgroup of their peers [27].

The probability distribution of degrees for Events partition is nothing more than a distribution of the event sizes. For all four social groups, this event size distribution follows log-normal behavior

$$P(q^E) = \frac{1}{q^E \sigma^E \sqrt{2\pi}} e^{-\frac{(\log(q^E)-\mu^E)^2}{2(\sigma^E)^2}}. \tag{17}$$

Three groups (GEAM, PGHF, and TECH) have the same value of mean $\mu^E = -0.7(1)$ and standard deviation $\sigma^E = 0.95(5)$, while the degree distribution for Events of LVHK group, with parameters $\sigma^E = 0.59(2)$ and $\mu^E = -0.16(2)$, shows some deviations from log-normal curve for small values of normalized degree. Log-normal distribution has been observed in many socioeconomic systems, including distribution of firm [41] and city [42] sizes, voting [11, 43] and citation patterns [12, 44]. For all these processes proportional growth is a common mechanism that leads to a log-normal distribution of event sizes. As it was argued in Section 2, the growth of Meetup groups can be described with multiplicative process [45]. Since event participation is the primary activity in a Meetup group, event size directly depends on the group's growth. Thus, the log-normal distribution of event sizes is just another indicator of the proportional growth of Meetup social groups.

We also examine the mixing patterns in these bipartite networks by calculating the average nearest neighbor degree for the nodes in both partitions and study its dependence on their degree, Fig. 4 (bottom). The average nearest neighbor degree for Members partition, i.e. the average size of events attended by a member, is independent of the magnitude of her activity Fig. 4(bottom right). The descending curve in Fig. 4(bottom left) indicates disassortative mixing for events. This observation is expected since in the previous study [27] we have shown that significant events are predominantly attended by members with a small number of participations.

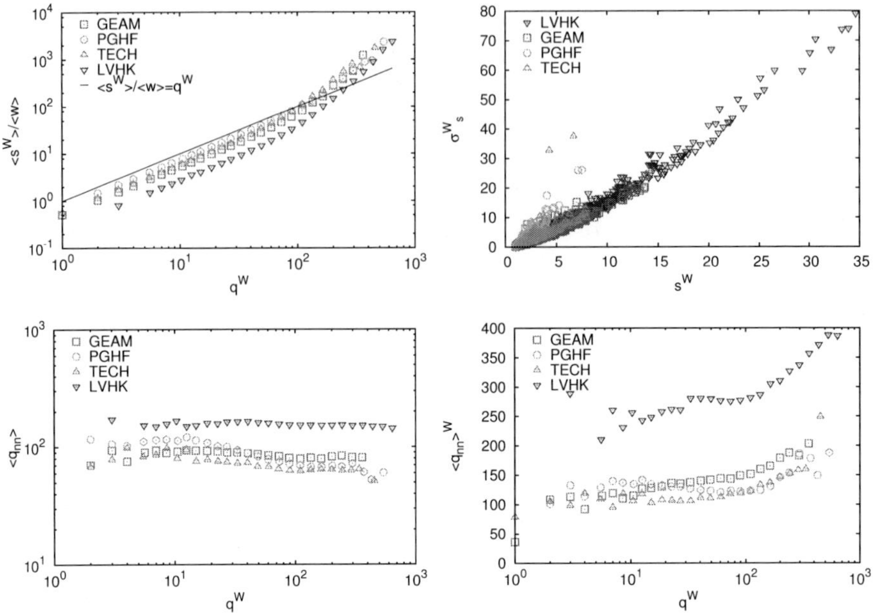

Figure 5. (top left) Dependence of normalized node strength on its degree. (top right) Scatter plot of nodes strength versus standard deviation of link weights adjacent to that node. Average nearest neighbor degree (bottom left) and its weighted counterpart (bottom right) in weighted network.

The observed degree-degree correlations suggest that event-driven dynamics is outlined with correlated social behavior which is beyond random processes characteristic for generation of random bipartite graphs. This event-driven dynamics of Meetup social group shapes and it is formed by the structure of the underlying weighted social network. Real world weighted networks, besides complex topology, also display a considerable heterogeneity in the strength of the connections. Weighted quantities and their correlations with the underlying topological structure of the network provide additional information about the dynamical processes governing network's evolution. We further explore topological features of the weighted network between the Meetup group members, which is obtained by projecting bipartite network onto Members partition and filtering out non-significant links.

Figure 5(left) shows dependence of normalized average strength $\langle s^W \rangle / \langle w \rangle$

of node on the value of its degree q^W. Normalization constant $\langle w \rangle$ is average link weight in the whole network. In the Meetup weighted social network the degree of a node equals to the number of different connections adjacent to that node, while its strength measures the total magnitude of these relationships. In uncorrelated weighted networks, the normalized average strength of the node grows with its degree as $\langle s^W \rangle / \langle w \rangle = q^W$. We see from 5(top left) that for all four groups the dependence of normalized average node's strength deviates from uncorrelated approximation. Furthermore, normalized average node's strength is proportional to its degree for nodes with degree $q^W \leq 50$, while for nodes with large values of degrees it grows super-linearly. This indicates that average weight of links adjacent to a node $\langle w_{i^M} \rangle = \frac{1}{q^M_{i^M_1}} \sum_{i^M_2} W_{i^M_1 i^M_2}$ grows with node degree. In fact, the standard deviation of weights of links adjacent to a node also increases with its strength, Fig. 5 (top right), which suggests that very active users are characterized by significant heterogeneity of strength of their social connections. The described dependencies reveal one of the essential characteristics of event-driven dynamics: at the beginning of their involvement in group activities members mostly widen their social circle, while later, as their participation in group activities progresses, they tend to pay more attention to strengthening their existing social relations with particular members.

It was shown above that mixing patterns for Members partition in the bipartite network of Meetup groups are uncorrelated, so it is not surprising that we observe the lack of degree-degree correlations in corresponding weighted social networks Fig. 5 (bottom left). On the other hand, the dependence of weighted average nearest neighbor degree $\langle q_{nn} \rangle^W$ on node degree shows weak assortative mixing. The weighted average nearest neighbor degree is defined in a way to put more emphasis on the properties of strongest connections. Thus we see that members with a large number of connections establish the most solid ones with other high-degree members, forming in this way strongly connected core of very active members. This finding is further confirmed with a comparison of non-weighted $C(q^W)$ and weighted clustering $C^W(q^W)$ clustering coefficient shown in Fig. 6. For all four groups, non-weighted clustering coefficient and its weighted counterpart have the same value for low- and middle-degree nodes. For nodes with $q^W \simeq 50$, the difference between these two topological measures becomes evident, and it further grows with the degree. The ratio between non-weighted and weighted clustering coefficient of very active users suggests that their involvement in group activities is not conditioned by the behaviour of

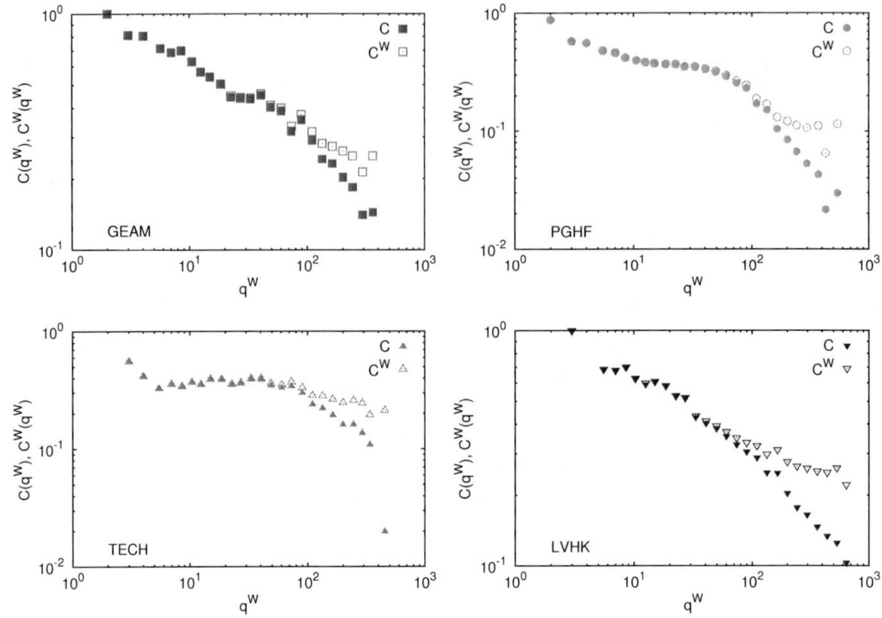

Figure 6. Dependence of non-weighted and weighted clustering coefficient on node degree for four studied Meetup social groups.

other members from their social circles but rather by the activity of the whole subgroups, as it was argued in [27].

This type of activity leads to the occurrence of communities which are manifested as mesoscopic inhomogeneities in the structure of weighted social networks, i.e. as the subgroups of strongly inter-connected nodes. Their existence, observed in many online and offline social systems [7,9,17,33], is one of the strongest indicators of self-organising collective dynamics. We detect these communities in the weighted social network of the 10% most active members using LOUVAIN community detection method [46,47] and its implementation in Gephi [48]. This approach detects communities in a network by finding the configuration of groups of nodes for which the modularity function has the highest value [46]. For all four groups we find that they have between four and five communities of similar sizes, see the Table 2 for details. For the illustration, we show the weighted social network of 10% most active individuals in LVHK and its community structure in Fig. 2(right).

Table 2. Details about weighted social subnetworks formed between 10% of most active members for considered networks. $\overline{N_W}$ and $\overline{L_W}$ are the number of members and links in subnetwork, activity range is the lowest and highest number of attended events by selected group of members, and N_C is the number of detected communities using LOUVAIN method

Acronym	$\overline{N_W}$	$\overline{L_W}$	Activity range	N_C
GEAM	538	15575	19-1494	4
PGHF	500	6828	15-1191	4
TECH	317	5350	21-1175	5
LVHK	607	29327	48-1521	4

All results presented in this section show that these four groups, although essentially different when it comes to their topic, type of activity and personality of their members, have very similar, practically universal, internal structure. This further confirms the conclusion from our previous research [26, 27] that the forces that influence the behavior of the individuals and thus govern the collective dynamics of the group are very fundamental, i.e. they do not depend on specific type of social group. In the next section we further support these findings by analysing group dynamics.

4. Dynamics of Meetup Social Groups

Specificity of Meetup social groups is their event-driven dynamics. Members of the group meet at a precise time and in a precisely determined place to participate in some joint activity. Through these activities, they form and strengthen social connections, which then influence the future dynamics of the group, i.e. the frequency and size of future events. The individual participations in group events can be translated into time series that either describes the activity of each member or the collective activity of the subgroup or the whole group. The quantitative analysis of these time series provides a characterization of system dynamics and reveals valuable information about its organization and evolution [7, 8, 10, 17, 33]. In this section, we analyze activity patterns of individual members as well as the total subgroup and group activity to find universal features of Meetup group dynamics.

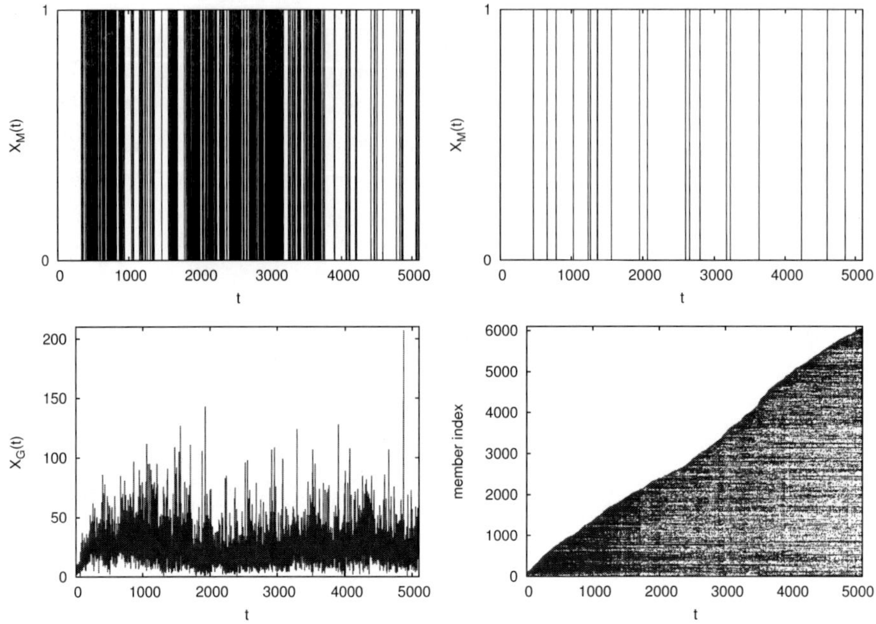

Figure 7. Time series of activity of one of the very active members (top left) and low active members (top right) of LVHK group. (bottom left) Time series of total activity in LVHK group. (bottom right) LVHK members activity pattern. The x axes in Fig.7 corresponds to event-time while y axes shows member's index. Members are ordered according to their first occurrence time in group activities, and dot denotes member's participation in event.

The number of organized events per week varies not just between different Meetup groups, but also within the group Fig. 1 (right). This frequency is influenced by various external and internal factors. To separate factors of different nature, we measure the activity in the group against the number of held events and thus omit real time from the further analysis. This way we obtain different time series $X(t)$ where t is a serial number of event. Time series of individual member's activity $X_{i^M}(t)$, shown in Fig. 7 (top), is an array of 0 and 1 (a member can either attend or not attend the event). The time series that describes the activity of the whole group $X_G(t)$ is obtained through aggregation of the time series for individual members. $X_G(t)$ is nothing else but the time series of event sizes Fig. 7(bottom left).

First, we study the patterns of member's participation in group's events. Based on time series of individual member's activity one can calculate the time lag between two attended events ΔT^{M}, measured in event-time, and obtain the time series of inter-event times for each member $\{\Delta T_1^{i^M}, \ldots, \Delta T_{q^M-1}^{i^M}\}$. As it was shown in [27] the distributions of these time interval for all four Meetup groups follow the truncated power-law behavior with power-law exponents in interval $[1.06, 1.38]$. This suggests that temporal patterns of members activity have fractal structure [8, 33] known as bursty behavior [49]. Bursty behavior, characterized by intervals with persistent activity followed by the long inactive intervals, is universal characteristic of human behavior and it has been observed in various social systems [22, 33, 50]. This type of activity results in particular pattern of members attendance of group events, shown for LVHK group in Fig. 7 (bottom right).

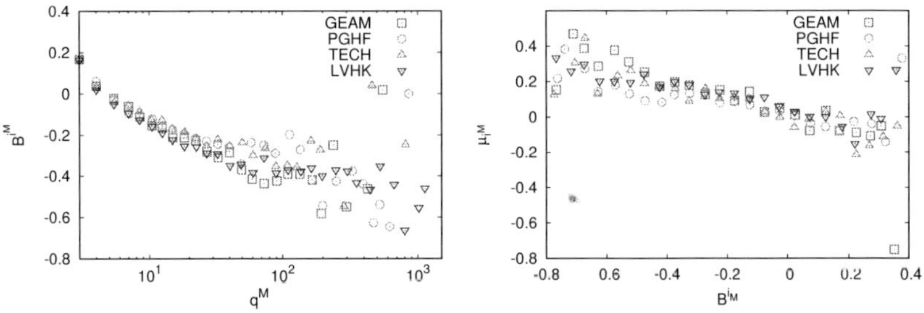

Figure 8. (left) Dependence of member's burstiness on the number of attended events. (right) Dependence of memory coefficient on burstiness.

It was shown in [49] that there are two different mechanisms responsible for the bursty nature of the signal: the inter-event time distribution and memory. Two measures used for distinguishing between these two mechanisms are burstiness parameter B and memory coefficient μ. The burstiness parameter is given by

$$B_{i^M} = \frac{\langle \Delta T^{i^M} \rangle - \sigma_{\Delta T^{i^M}}}{\langle \Delta T^{i^M} \rangle + \sigma_{\Delta T^{i^M}}}, \tag{18}$$

where $\langle \Delta T^{i^M} \rangle$ and $\sigma_{\Delta T^{i^M}}$ are average value and standard deviation of time lag between two attended events respectively. B_{i^M} takes values between -1 and

1 and it is used for quantifying the inter-event time distribution of the signal: $B_{i^M} = 1$ corresponds to very bursty signal, $B_{i^M} = 0$ to random signal, and $B_{i^M} = -1$ corresponds to very regular (periodic) signal. The memory coefficient μ, a correlation-based measure, measures autocorrelations between consecutive inter-event times $\left(\Delta T_k^{i^M}, \Delta T_{k+1}^{i^M}\right)$ for one member. It is calculated as

$$\mu_{i^M} = \frac{1}{q^M - 2} \sum_{k=1}^{q^M-2} \frac{(\Delta T_k^{i^M} - \langle \Delta T^{i^M} \rangle_1)(\Delta T_{k+1}^{i^M} - \langle \Delta T^{i^M} \rangle_2)}{\sigma_1 \sigma_2}, \tag{19}$$

where $\langle \Delta T^{i^M} \rangle_1$ $\left(\langle \Delta T^{i^M} \rangle_2\right)$ and $\sigma_1(\sigma_2)$ are sample mean and sample standard deviation of $\Delta T_k^{i^M}$ $\left(\Delta T_{k+1}^{i^M}\right)$ respectively.

We quantify the temporal patterns of individual member's activities by calculating the burstiness parameter B_{i^M} and memory coefficient μ_{i^M} [49]. Figure 8 (left) shows how burstiness of individual members depends on the size of their activity. Figure 8 indicates that burstiness parameter decreases as the involvement of member in group grows, suggesting that very active members attend events on more regular basis than members with a small or moderate number of participations. This is in line with previous findings [26, 27] where it was shown that probability of members future involvement in group activities depends nonlinearly on the ratio between the number of previous participations and non-participations in group events. Specifically, for very active members their probability to attend the next event is very close to one and does not vary much with time, which results in relatively regular patterns of participation (see Fig. 7 (top left)). On the other hand, a member with a few attendances has the typical pattern shown in Fig. 7(top right): inter-event time distribution is more a Poissonian like with burstiness parameter close to zero. The dependence of memory coefficient μ_{i^M} on burstiness parameter, shown in Fig. 8 (right), further confirms that members probability of attending the next group event depends on the number of previously attended and not-attended events. Members with the largest number of attended events, i.e. a negative value of burstiness parameter, have the highest and positive value of memory coefficient. This suggests that the origin of burstiness of their activity is memory and that regular members have more predictable group activity patterns than temporary members. These findings show that human event-participation dynamics is very different from one typically observed in social systems where human behavior is characterized with high burstiness parameter and memory coefficient close to zero.

As we saw in Section 3, members with medium and high number of atten-

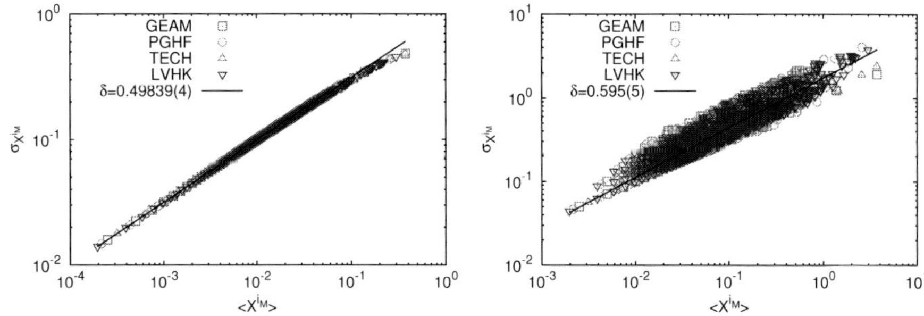

Figure 9. The relationship between fluctuations $\sigma_{X_{iM}}$ and average $\langle X_{iM} \rangle$ activity of member time series at the scale of one (left) and ten events (right).

dance have large circles of acquaintances within which we can distinguish subgroups of strongly connected friends. It was argued in [27] that these connections and member's need to maintain them are the primary drivers behind her involvement in group activities. On the other hand, it was shown that dynamics of various social groups is externally driven by the arrival of new members [9, 17]. We use formalism introduced in [51] to determine in quantitative manner which of the factors, endogenous or exogenous, are responsible for Meetup group dynamics. Based on time series of activity, for every member, we calculate time average activity $\langle X_{iM} \rangle$ and standard deviation $\sigma_{X_{iM}}$. The analysis of dynamics in various systems has shown that relationship between standard deviation and time average of activity follows the scaling law

$$\sigma_{X^{iM}} \propto \langle X_{iM} \rangle^{\delta} . \tag{20}$$

The value of exponent $\delta \approx 0.5$ corresponds to system where endogenous factors dominate group dynamics, $\delta \approx 1$ to system driven by external forces, while the values of δ between these two extremal values denote mixed dynamics. Figure 9 (left) shows a perfect scaling for all four groups with value of exponent $\delta \approx 0.5$, which indicates that Meetup group dynamics is practically random on the time scale of one event. If we do the same analysis on the time series obtained by aggregating members activity over ten events (which corresponds to one week in real time), we see that the scaling exponent δ becomes larger than 0.5 suggesting correlated members activity at longer time scales, Fig. 9 (right).

The community structure of the network of the 10% most active members is a direct consequence of correlated co-occurrence of members at group events.

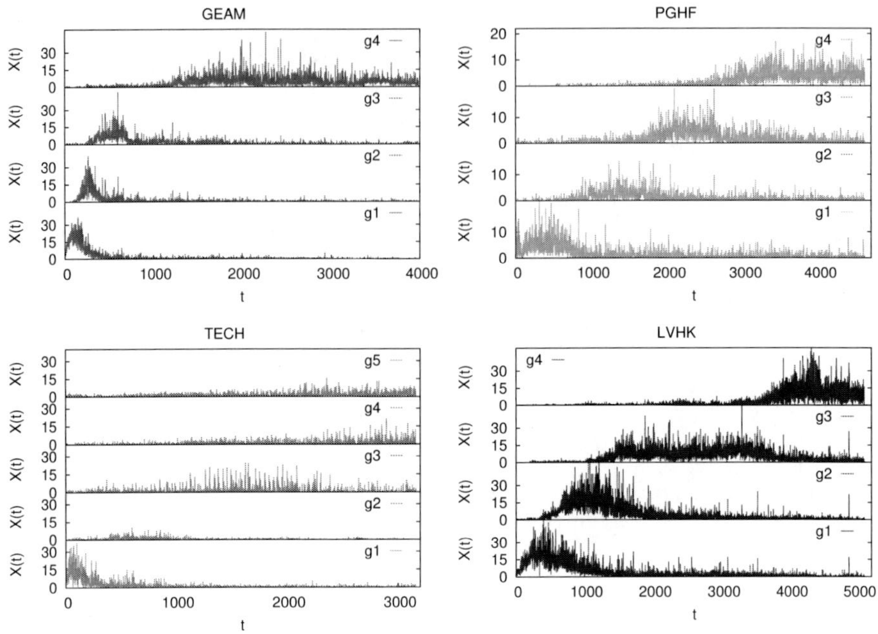

Figure 10. Total activity of subgroups found in weighted networks of the 10% most active members using LOUVAIN method.

Figure 10 shows the temporal evolution of activity in these communities. The time series shown in Fig. 10 are obtained by aggregating time series of individual actions of all members belonging to one community. We see that all these communities have similar paths of evolution: at the beginning, the activity of their members is low, almost sporadic; then we observe a gradual growth, followed by the peak of community activity, after which the activity decreases until the community eventually ceases to exist. This rise and fall of social groups have been observed in both online and offline social systems [8, 9, 52]. The ability of group to dynamically alternate their composition is one of the main features of persistent large groups. Obviously, the activity of members belonging to one community is strongly correlated; we see that there is only small overlap between the activity of different communities. This further indicates that significant social relations are clustered in time and that these relations are main driving force behind the dynamics of Meetup social groups [26, 27].

Conclusion

Data from Meetup groups contain invaluable information about the human event-driven offline activity. We have demonstrated how the combination of computational methods from complex network theory and statistical physics can be used to explore the structure and dynamics of four different Meetup social groups. Specifically, we have shown how the data from these networks can be mapped onto bipartite networks of Members and Events, and how to project these networks onto Members partition to obtain social relations. These networks have several specific features that are universal for Meetup and, presumably other event-driven, social networks. In particular, both network types have very rich topology which is directly related to event-driven group dynamics. Bipartite networks are characterized by high degree heterogeneity in both partitions, which is not surprising, since power-law dependencies are the distinctive feature of members attendance patterns in event-driven social groups [26, 27]. What is surprising is a lack of degree-degree correlations in Members community, which suggests that members attend events of different sizes with equal probability. The weak degree-degree correlations observed for Events partition indicate that small events are predominately attended by active members of Meetup group, which confirms the findings of event size importance in [27]. The analysis of features of weighted networks of significant social links shows that network topology alone is not sufficient for capturing and understanding the social mechanism that govern the Meetup group dynamics and that link weights and their distributions need to be taken into account. The analysis of weighted measures shows that connection weights become important for dynamics of members with more than 50 connections. It also indicates that very active members tend to form strongly connected subgroups. The communities, identified as topological mesoscopic inhomogeneities in weighted social networks of the 10% most active members, are distinctive sign that these groups have self-organizing dynamics.

Meetup group collective dynamics is shaped by the behavior of individual member and interactions between them. The analysis of patterns of members behavior has again shown that human dynamics is bursty. What is different in the case of Meetup dynamics is that origin of this burstiness is memory, which is different from what is observed in other social dynamics. The positive value of memory coefficient for active users shows that their activity is correlated in time. We have shown that this correlation can be observed at the scale of ten and more event, which corresponds to one week in real time. This correlated behavior is also the principal mechanism which leads to formation of communities in social network of the 10% most active users.

It has been shown earlier that Meetup group dynamics is universal, i.e. does not depend on group category, location or type of activity [27]. The comparison of this dynamics with conference participation patterns indicates that this universality is independent of members motivation to take part in groups activities, for instance, professional versus leisure. Here we show that this universality is also observed in the structure of social networks. Members of all four Meetup groups form networks which have a remarkably similar structure: they all exhibit similar degree distributions, have the same degree-degree correlations, and very similar community structure with similar number of communities.

The methodology that we demonstrated here by applying it to Meetup social groups is a standard part of social computational science, an emerging interdisciplinary scientific field. The presented methods and tools can be used in study of other social networks whose activity is localized in time and space.

Acknowledgment

This work was supported by the Ministry of Education, Science, and Technological Development of the Republic of Serbia under project ON171017. Numerical simulations were performed on the PARADOX supercomputing facility at the Scientific Computing Laboratory of the Institute of Physics Belgrade.

References

[1] Lazer D, Pentland AS, Adamic L, Aral S, Barabasi AL, Brewer D, et al. Life in the network: the coming age of computational social science. *Science*. 2009; 323(5915):721.

[2] Conte R, Gilbert N, Bonelli G, Cioffi-Revilla C, Deffuant G, Kertesz J, et al. Manifesto of computational social science. *European Physical Journal-Special Topics*. 2012; 214:325.

[3] Castellano C, Fortunato S, Loreto V. Statistical physics of social dynamics. *Rev Mod Phys*. 2009 May;81:591–646.

[4] Sen P, Chakrabarti BK. *Sociophysics: an introduction*. Oxford University Press; 2013.

[5] Schelling TC. Dynamic models of segregation. *Journal of mathematical sociology.* 1971; 1(2):143–186.

[6] Putterman L. Cooperation and punishment. *Science.* 2010; 328(5978):578–579.

[7] Mitrović M, Paltoglou G, Tadić B. Networks and emotion-driven user communities at popular blogs. *The European Physical Journal B-Condensed Matter and Complex Systems.* 2010; 77(4):597–609.

[8] Mitrović M, Paltoglou G, Tadić B. Quantitative analysis of bloggers' collective behavior powered by emotions. *Journal of Statistical Mechanics: Theory and Experiment.* 2011; 2011(02):P02005. Available from: http://stacks.iop.org/1742-5468/2011/i=02/a=P02005.

[9] Mitrović M, Tadić B. Dynamics of bloggers' communities: bipartite networks from empirical data and agent-based modeling. *Physica A: Statistical Mechanics and its Applications.* 2012;391(21):5264–5278.

[10] Šuvakov M, Mitrović M, Gligorijević V, Tadić B. How the online social networks are used: dialogues-based structure of MySpace. *Journal of The Royal Society Interface.* 2013;10(79):20120819.

[11] Chatterjee A, Mitrović M, Fortunato S. Universality in voting behavior: an empirical analysis. *Scientific Reports.* 2013;3:1049.

[12] Radicchi F, Fortunato S, Castellano C. Universality of citation distributions: Toward an objective measure of scientific impact. *Proceedings of the National Academy of Sciences.* 2008;105(45):17268–17272.

[13] Dunbar RIM. Coevolution of neocortical size, group size and language in humans. *Behavioral and Brain Sciences.* 1993 12;16(4):681–694.

[14] Hill RA, Dunbar RIM. Social network size in humans. *Human Nature.* 2003;14(1):53–72.

[15] Dunbar RIM. Mind the gap: or why humans aren't just great apes. *Proceedings of the British Academy.* 2008;154:403–423.

[16] Dunbar RIM. Constraints on the evolution of social institutions and their implications for information flow. *Journal of Institutional Economics.* 2011;7(3):345–371.

[17] Mitrović Dankulov M, Melnik R, Tadić B. The dynamics of meaningful social interactions and the emergence of collective knowledge. *Scientific Reports*. 2015;5:12197.

[18] Aral S, Walker D. Identifying influential and susceptible members of social networks. *Science*. 2012;337(6092):337–341.

[19] González-Bailán S, Borge-Holthoefer J, Moreno Y. Broadcasters and hidden influentials in online protest diffusion. *American Behavioral Scientist*. 2013;57(7):943–965.

[20] Lin YR, Chi Y, Zhu S, Sundaram H, Tseng BL. Facetnet: a framework for analyzing communities and their evolutions in dynamic networks. In: *Proceedings of the 17th International Conference on World Wide Web*. WWW '08; 2008. p. 685–694.

[21] Garas A, Garcia D, Skowron M, Schweitzer F. Emotional persistence in online chatting communities. *Scientific Reports*. 2012;2:402.

[22] Yasseri T, Sumi R, Rung A, Kornai A, Kertsz J. Dynamics of conflicts in wikipedia. *PLoS ONE*. 2012 06;7(6):1–12.

[23] Montazeri A, Jarvandi S Soghraand Haghighat, Vahdani A Mariamand Sajadian, Ebrahimi M, Haji-Mahmoodi M. Anxiety and depression in breast cancer patients before and after participation in a cancer support group. *Patient education and counseling*. 2001;45:195–198.

[24] Davison KP, Pennebaker JW, Dickerson SS. Who talks? The social psychology of illness support groups. *American Psychologist*. 2000;55:205–217.

[25] Tam Cho WK, Gimpel JG, Shaw DR. The tea party movement and the geography of collective action. *Quarterly Journal of Political Science*. 2012;7:105–133.

[26] Smiljanić J, Chatterjee A, Kauppinen T, Mitrović Dankulov M. A theoretical model for the associative nature of conference participation. *PLoS ONE*. 2016 02;11(2):1–12.

[27] Smiljanić J, Dankulov MM. Associative nature of event participation dynamics: A network theory approach. *PloS one*. 2017;12(2):e0171565.

[28] Meetup;. Accessed: 2017-06-30. https://www.meetup.com/.

[29] Meetup Dataset;. Accessed: 2016-06-27. https://figshare.com/articles/Meetup_Datasets/2066904.

[30] Qiao Z, Zhang P, Zhou C, Cao Y, Guo L, Zhang Y. Event recommendation in event-based social Networks. In: *Proceedings of the Twenty-Eighth AAAI Conference on Artificial Intelligence.* AAAI'14; 2014. p. 3130–3131.

[31] Zhang W, Wang J, Feng W. Combining latent factor model with location features for event-based group recommendation. In: *Proceedings of the 19th ACM SIGKDD International Conference on Knowledge Discovery and Data Mining.* KDD '13; 2013. p. 910–918.

[32] Liu X, He Q, Tian Y, Lee WC, McPherson J, Han J. Event based social networks: linking the online and offline social worlds. In: *Proceedings of the 18th ACM SIGKDD International Conference on Knowledge Discovery and Data Mining.* KDD '12; 2012. p. 1032–1040.

[33] Mitrović M, Tadić B. Bloggers behavior and emergent communities in Blog space. *Eur Phys J B.* 2010;73(2):293–301.

[34] Dianati N. Unwinding the hairball graph: Pruning algorithms for weighted complex networks. *Physical Review E.* 2016 Jan;93:012304.

[35] Saracco F, Di Clemente R, Gabrielli A, Squartini T. Randomizing bipartite networks: the case of the World Trade Web. *Scientific Reports.* 2015;5:10595.

[36] Saracco F, Straka MJ, Di Clemente R, Gabrielli A, Caldarelli G, Squartini T. Inferring monopartite projections of bipartite networks: an entropy-based approach. *New Journal of Physics.* 2017;19(5):053022.

[37] Bianconi G, Coolen ACC, Perez Vicente C. Entropies of complex networks with hierarchically constrained topologies. *Physical Review E.* 2008;78(1):016114.

[38] Anand K, Bianconi G. Entropy Measures for Networks: Toward an Information Theory of Complex Topologies. *Phys Rev E.* 2009;80:045102(R).

[39] Boccaletti S, Latora V, Moreno Y, Chavez M, Hwang DU. Complex networks: structure and dynamics. *Physics reports*. 2006;424(45):175 – 308.

[40] Barrat A, Barthlemy M, Pastor-Satorras R, Vespignani A. The architecture of complex weighted networks. *Proceedings of the National Academy of Sciences of the United States of America*. 2004;101(11):3747–3752.

[41] Stanley MH, Buldyrev SV, Havlin S, Mantegna RN, Salinger MA, Stanley HE. Zipf plots and the size distribution of firms. *Economics letters*. 1995;49(4):453–457.

[42] Gabaix X, Ioannides YM. The evolution of city size distributions. *Handbook of regional and urban economics*. 2004;4:2341–2378.

[43] Fortunato S, Castellano C. Scaling and universality in proportional elections. *Physical Review Letters*. 2007;99(13):138701.

[44] Chatterjee A, Ghosh A, Chakrabarti BK. Universality of citation distributions for academic institutions and journals. *PloS one*. 2016;11(1):e0146762.

[45] Mitzenmacher M. A brief history of generative models for power law and lognormal distributions. *Internet mathematics*. 2004;1(2):226–251.

[46] Blondel VD, Guillaume JL, Lambiotte R, Lefebvre E. Fast unfolding of communities in large networks. *Journal of statistical mechanics: theory and experiment*. 2008;2008(10):P10008.

[47] Lambiotte R, Delvenne JC, Barahona M. *Laplacian dynamics and multiscale modular structure in networks*. arXiv preprint arXiv:08121770. 2008;.

[48] Bastian M, Heymann S, Jacomy M. *Gephi: an open source software for exploring and manipulating networks*. 2009;.

[49] Goh KI, Barabási AL. Burstiness and memory in complex systems. *Europhysics Letters*. 2008;81(4):48002.

[50] Karsai M, Kaski K, Barabási AL, Kertész J. Universal features of correlated bursty behaviour. *Scientific Reports*. 2012;2.

[51] De Menezes MA, Barabási AL. Fluctuations in network dynamics. *Physical Review Letters*. 2004;92(2):028701.

[52] Palla G, Barabási AL, Vicsek T. Quantifying social group evolution. *Nature*. 2007;446(7136):664 667.

In: Scientific Computing
Editor: Caj Erling
ISBN: 978-1-53612-564-1

Chapter 3

EFFICIENT NUMERICAL TOOLS FOR SOLVING THE NONLINEAR SCHRÖDINGER EQUATION

Vladimir Lončar, Ivana Vasić and Antun Balaž*
Scientific Computing Laboratory,
Center for the Study of Complex Systems,
Institute of Physics Belgrade, University of Belgrade,
Belgrade, Serbia

Abstract

The nonlinear Schrödinger equation or Gross-Pitaevskii equation plays an important role in several areas of physical sciences. It is used in various forms to describe prominent physical phenomena, such as propagation of light in a nonlinear medium and the dynamics of weakly interacting Bose-Einstein condensates studied in experiments with ultracold quantum gases. Large-scale numerical simulations for realistically large physical systems based on this equation have provided valuable insights into the properties of many interesting phenomena. Furthermore, numerical solutions of this equation are now routinely used for interpretation of experimental data and are indispensable for day-to-day operation of many laboratories, as well as for the design of new experimental setups.

In this chapter we review an efficient numerical algorithm for solving this nonlinear partial differential equation in three spatial coordinates

*Corresponding Author Email: antun.balaz@ipb.ac.rs.

and either real or imaginary time, based on the Crank-Nicolson split-step semi-implicit method. Integration over imaginary time provides a direct access to stationary solutions of the full time-dependent problem. General spatially anisotropic solutions in three spatial variables are considered, as well as dimensionally-reduced problems such as radially or spherically symmetric cases.

Stability and efficiency of the presented method is ensured in the following way. The discretization error in both time and spatial step is of the second order. Moreover, integration over the time variable is split into two steps that are dealt with in a different, but optimized way. In the first step, only local part of the equation is taken into account and integration is performed exactly with respect to it. In this way the nonlinear term is considered in a straightforward, but optimized way that allows for a treatment of strong nonlinearities. In the second step, spatial derivatives are considered separately. Semi-implicit nature of the algorithm also contributes to its stability.

This method is numerically implemented using the standard C programming language. Further computational optimization is achieved by employing the state-of-the-art parallelization techniques for different computing platforms: OpenMP (for shared memory multiprocessing), OpenMP/MPI (for distributed memory systems, where each compute node uses OpenMP to maximize performance), CUDA (for single GPGPUs), CUDA/MPI (for distributed memory systems, where each compute node has GPGPU installed), as well as their hybrid combinations. The scalability of the algorithm with the number of processors and compute nodes is discussed.

As applications and extensions of the above developed algorithms, we present three relevant examples from the field of ultracold atoms. The first application illustrates how the non-equilibrium dynamics of multicomponent Bose-Einstein condensates can be numerically studied, including the excitation dynamics of a condensate with a coupling of the effective spin and angular momentum in the second application. The third application demonstrates how formation of vortices can be studied in a condensate featuring contact, as well as long-range dipole-dipole interaction.

PACS: 02.60.Lj, 02.60.Jh, 02.60.Cb, 03.75.-b

Keywords: Bose-Einstein condensate, Gross-Pitaevskii equation, Dipole-dipole interaction, OpenMP, MPI, CUDA, Crank-Nicolson method

1. Introduction

One of the important breakthroughs in physics of the twentieth century was the development of quantum mechanics. The key ingredient of this theory is the Schrödinger equation: a linear equation that governs the time evolution of a quantum system. In one of its common representations, Schrödinger equation is a partial differential equation.

Full quantum description of an interacting many–body system easily becomes computationally demanding due to exponentially increasing underlying vector space with the system size. For this reason, approximate descriptions that capture main physical phenomena are highly sought–after. The nonlinear Schrödinger (NLS) equation represents an effective, mean–field description of a quantum system. It is of great relevance in the field of cold atoms, in particular for a description and understanding of a weakly interacting bosonic systems at low temperatures in the regime of Bose–Einstein condensation [1]. In this context, the equation is widely known as Gross–Pitaevskii equation [2, 3] and it has provided valuable theoretical understanding of experimental results. Another field of research where the NLS equation plays a prominent role is the field of nonlinear optics that was initiated by the discovery of lasers in 1960 [4]. As analytical solutions of the NLS equation are limited to just a few special cases, the NLS equation was investigated numerically starting in the seventies using several different numerical methods [5]. The development of present–day computer architectures promotes numerical simulations and enables studies of more complex physical problems for experimentally relevant system sizes.

In this chapter we first introduce the NLS equation and derive an algorithm for obtaining its solutions based on a split–step method. In the next step we present an optimized implementation of this algorithm that heavily exploits parallelization techniques available on present–day state–of–the–art computer architectures. In the last sections of this chapter we illustrate applications of the described method for three examples that are currently in a research focus of cold–atom community.

2. Algorithm and Computational Implementation

2.1. Introduction

The time–dependent nonlinear Schrödinger equation in a dimensionless form is given by

$$i\frac{\partial\psi(\mathbf{r},t)}{\partial t} = \left(-\frac{1}{2}\Delta + V(\mathbf{r}) + g|\psi(\mathbf{r},t)|^2\right)\psi(\mathbf{r},t), \quad (1)$$

where Δ is the three-dimensional Laplace operator $\Delta \equiv \frac{\partial^2}{\partial x^2} + \frac{\partial^2}{\partial y^2} + \frac{\partial^2}{\partial z^2}$. The usual Schrödinger equation describing the time–evolution of a single–particle wave function $\psi(\mathbf{r},t)$ is recovered by eliminating the nonlinear term by setting $g = 0$. The wave function depends on four variables: three spatial coordinates $\mathbf{r} = (x, y, z)$ and time t. The real–valued function $V(\mathbf{r})$ corresponds to an external potential, and before mentioned Laplacian stems from the kinetic energy. Initial conditions are appropriately set by fixing $\psi(\mathbf{r}, t = 0) = \psi_{\text{initial}}(\mathbf{r})$ and then the time evolution for $t > 0$ is studied. The equation (1) conserves the normalization of the function $\psi(\mathbf{r},t)$

$$\int d\mathbf{r}\, |\psi(\mathbf{r},t)|^2 = 1\,, \quad (2)$$

which in physical terms corresponds to the conservation of the particle number, as well as the total energy of the system,

$$E = \int d\mathbf{r}\, \psi^*(\mathbf{r},t)\left(-\frac{1}{2}\Delta + V(\mathbf{r}) + \frac{g}{2}|\psi(\mathbf{r},t)|^2\right)\psi(\mathbf{r},t)\,. \quad (3)$$

In the context of cold atoms, the complex–valued function $\psi(\mathbf{r},t)$ stands for a condensate wave function – an effective single particle wave function macroscopically occupied by bosonic particles. The constant g is set by weak atomic interactions that we approximate by contact interactions: repulsive interactions yield $g > 0$, while attractive interactions are described by $g < 0$. The phase of a system described by the nonlinear Schrödinger equation is called a Bose–Einstein condensate (BEC). Macroscopic occupation of a single particle state (a finite condensate fraction) is a consequence of bosonic statistics (Bose–Einstein statistics) that becomes of key importance at very low temperatures. The phenomenon of Bose–Einstein condensation was predicted in early days of quantum mechanics in 1924 when the Bose–Einstein statistics was derived for the first time. After several decades of experimental efforts, first experimental realization of a BEC with nanokelvin-cold atoms was reported in 1995 [6, 7, 8]. This success marked the onset of intense research activity in the field of cold atoms where the description based on the nonlinear Schrödinger equation played an important role [9, 10, 11, 12, 1, 13, 14]. Formally, the equation (1) can be derived by using the time–dependent variational principle. The time–dependent spatial density distribution of a condensate is given by

$$n(\mathbf{r},t) = |\psi(\mathbf{r},t)|^2 . \tag{4}$$

Equation (1) governs the time evolution of a weakly interacting BEC. Another relevant question concerns finding a stationary solution $\psi(\mathbf{r})$ that minimizes the energy

$$E_0 = \int d\mathbf{r}\, \psi^*(\mathbf{r}) \left(-\frac{1}{2}\Delta + V(\mathbf{r}) + \frac{g}{2}|\psi(\mathbf{r})|^2 \right) \psi(\mathbf{r}) , \tag{5}$$

for a given potential $V(\mathbf{r})$ and interaction constant g. For a moment we are considering a stationary problem, so ψ is a function of three spatial variables only. To address this question we will use a variant of a well established gradient descent method [15]. To this end, we introduce a label $n = 0, 1, \ldots$, where $\psi(\mathbf{r})_0$ is our initial guess for the stationary solution. In the next step, the function ψ is incrementally changed as

$$\psi(\mathbf{r})_{n+1} - \psi(\mathbf{r})_n = -\epsilon \frac{\delta E_0(\psi)}{\delta \psi^*} , \tag{6}$$

where ϵ is an optimally chosen update step. The reasoning behind equation (6) is a general idea of the gradient descent method: we move through the space of functions $\psi(\mathbf{r})$ along the direction in which the energy functional exhibits the fastest decay. For a stationary solution $\psi_0(\mathbf{r})$ it holds true $\frac{\delta E_0(\psi)}{\delta \psi^*} = 0$ and it corresponds to an extremum (minimum) of E_0. The last equation can be rewritten as

$$\frac{\partial \psi(\mathbf{r},\tau)}{\partial \tau} = -\left(-\frac{1}{2}\Delta + V(\mathbf{r}) + g|\psi(\mathbf{r},\tau)|^2 \right) \psi(\mathbf{r},\tau) . \tag{7}$$

where $\tau = n\epsilon$ is a label that keeps track of propagation through the space of functions $\psi(\mathbf{r})$. We notice that equation (7) can be derived from equation (1) by a formal replacement $t \rightarrow i\tau$. For this reason, the variable τ is named imaginary time. In contrast to the energy conservation (3), imaginary time propagation by its construction ensures that with a right choice of initial function, we will reach a minimum of the energy functional (5) at long enough τ. Moreover, imaginary time propagation as given in (7) does not preserve normalization (2), so this condition should be provided additionally, either by introducing an appropriate Lagrange multiplier or by re–normalizing $\psi(\mathbf{r},\tau)$ in each propagation step (6). Finally, we note that the solution $\psi_0(\mathbf{r})$ which minimizes E_0 evolves in time according to the nonlinear equation (1) as

$$\psi(\mathbf{r},t) = e^{-i\mu t}\psi_0(\mathbf{r})\,, \tag{8}$$

where μ is a chemical potential

$$\mu = \int d\mathbf{r}\,\psi_0^*(\mathbf{r})\left(-\frac{1}{2}\Delta + V(\mathbf{r}) + g|\psi_0(\mathbf{r})|^2\right)\psi_0(\mathbf{r})\,. \tag{9}$$

More recent experimental progress in the field of cold atoms has enabled study of more complex and fascinating physical systems whose theoretical understanding necessitates generalizations and extensions of the basic NLS equation (1). One of the topics that has attracted lot of attention recently is realization of a BEC of atoms with strong magnetic dipole moment [16, 17, 18]. To describe the system, we define z–axis as the polarization axis of the dipole moments. In this case the dipole–dipole interaction is proportional to [19]

$$V_{\mathrm{dd}}(\mathbf{R}) = \frac{1-3\cos^2\theta}{|\mathbf{R}|^3}\,, \tag{10}$$

where vector $\mathbf{R} = \mathbf{r} - \mathbf{r}'$ stands for the relative position of the two dipoles, and θ is the angle between the vector $\mathbf{R}$ and the z–axis. The dipolar NLS equation is given by

$$i\frac{\partial\psi(\mathbf{r},t)}{\partial t} = \left(-\frac{1}{2}\Delta + V(\mathbf{r}) + g|\psi(\mathbf{r},t)|^2 + g_{\mathrm{dd}}\int d\mathbf{r}' V_{\mathrm{dd}}^{3D}(\mathbf{r}-\mathbf{r}')|\psi(\mathbf{r}',t))|^2\right)\psi(\mathbf{r},t). \tag{11}$$

The coefficient g_{dd} sets the strength of the dipolar interaction. Other generalizations of the NLS equation (1) will be discussed in sections 3 (two–component BEC) and 4 (BEC with the coupling of spin and angular momentum).

In order to further illustrate wide applicability of the NLS equation, here we present its form used in nonlinear fiber optics. Starting with Maxwell's equations and using several justified approximations, it can be shown that the propagation of the pulse filed envelope $A(x, y, t, z)$ along the z direction is governed by the NLS equation of the form [20, 4]

$$i\frac{\partial A(x,y,t,z)}{\partial z} = \left[-\frac{1}{2}\left(\frac{\partial^2}{\partial x^2} + \frac{\partial^2}{\partial y^2} + \beta\frac{\partial^2}{\partial t^2}\right) + \gamma|A(x,y,t,z)|^2\right] A(x,y,t,z)\,, \quad (12)$$

where the coefficient β describes dispersive effects and the coefficient γ describes nonlinear effects. In the following, we mainly consider equations (1) and (11), but we note that all the discussed algorithms and their numerical implementations can be used to simulate equation (12) as well.

2.2. Algorithm

As the NLS equation (1) has been widely used and solved numerically, various types of numerical methods have been applied to this purpose [5]. A recent review paper [21] provides extensive comparison of the properties of different numerical methods used to solve the NLS equation. Broadly speaking, numerical methods for solving equation (1) are either finite difference approximations, pseudo spectral methods or finite element methods. Depending on the applied time–discretization scheme, a finite difference method can be explicit, implicit, semi–implicit or split–step approximation. Here we follow derivations from reference [22, 23] and present a simple and efficient numerical method to integrate the NLS equation (1). From the numerical point of view, real–time propagation according to equation (1) and imaginary–time propagation (7) can be implemented in a very similar manner.

In order to integrate equations (1) and (7) in time, we perform time discretization $t \equiv t_0 + n\varepsilon$, $n = 0, 1, 2, \ldots$, $\psi(\mathbf{r}, t) \to \psi(\mathbf{r})_n$. It turns out that optimal ways to deal with spatially local and non–local terms of these equations are different, hence we further split integration over time into two consecutive pieces. This approximation formally corresponds to a replacement of equation (1) by a set of two equations

$$i\frac{\partial \psi(\mathbf{r},t)}{\partial t} = \left(V(\mathbf{r}) + g|\psi(\mathbf{r},t)|^2\right)\psi(\mathbf{r},t) \quad (13)$$

$$i\frac{\partial \psi(\mathbf{r},t)}{\partial t} = -\frac{1}{2}\Delta\psi(\mathbf{r},t), \quad (14)$$

assuming that the result of integration of equation (13) is the initial condition for solving equation (14). The split-step approximation given by (13) and (14) is valid only for short-enough time interval ε and introduces a numerical error of the order $O(\varepsilon^2)$.

For practical reasons in a general three-dimensional situation, we further split equation (14) into three parts treating spatial derivatives over the three spatial coordinates separately. Thus, in this way we obtain a set of four consecutive equations

$$i\frac{\partial\psi(\mathbf{r},t)}{\partial t} = H_i\,\psi(\mathbf{r},t) \tag{15}$$

where $i \in \{1,2,3,4\}$ and we introduced

$$H_1 = V(\mathbf{r}) + g|\psi(\mathbf{r},t)|^2, \tag{16}$$

$$H_2 = -\frac{1}{2}\frac{\partial^2}{\partial x^2}, \quad H_3 = -\frac{1}{2}\frac{\partial^2}{\partial y^2}, \quad H_4 = -\frac{1}{2}\frac{\partial^2}{\partial z^2}. \tag{17}$$

Accordingly, between configurations n and $n+1$, we introduce three intermediate states labeled by $n+\frac{1}{4}, n+\frac{2}{4}, n+\frac{3}{4}$ such that

$$\psi(\mathbf{r})_{n+\frac{1}{4}} = \exp\left[-i\varepsilon\left(V(\mathbf{r}) + g|\psi(\mathbf{r})_n|^2\right)\right]\psi(\mathbf{r})_n\,, \tag{18}$$

and

$$\psi(\mathbf{r})_{n+\frac{2}{4}} - \psi(\mathbf{r})_{n+\frac{1}{4}} = \frac{i\varepsilon}{4}\frac{\partial^2}{\partial x^2}\left[\psi(\mathbf{r})_{n+\frac{2}{4}} + \psi(\mathbf{r})_{n+\frac{1}{4}}\right], \tag{19}$$

$$\psi(\mathbf{r})_{n+\frac{3}{4}} - \psi(\mathbf{r})_{n+\frac{2}{4}} = \frac{i\varepsilon}{4}\frac{\partial^2}{\partial y^2}\left[\psi(\mathbf{r})_{n+\frac{3}{4}} + \psi(\mathbf{r})_{n+\frac{2}{4}}\right], \tag{20}$$

$$\psi(\mathbf{r})_{n+1} - \psi(\mathbf{r})_{n+\frac{3}{4}} = \frac{i\varepsilon}{4}\frac{\partial^2}{\partial z^2}\left[\psi(\mathbf{r})_{n+1} + \psi(\mathbf{r})_{n+\frac{3}{4}}\right]. \tag{21}$$

The result (18) gives an exact, analytical solution of equation (13). In this way the nonlinear term has been properly taken into account providing a stability of the employed numerical scheme for arbitrary strong nonlinearity coefficient g. In the equations (19, 20, 21), we have used the semi–implicit Crank–Nicolson method [24], which is second–order accurate in the time step ε and unconditionally stable method. The accuracy of the approximation is $O(\varepsilon^2)$.

To reduce the obtained differential equations to the algebraic form, we additionally perform space discretizations with the discretization steps h_x, h_y, and h_z. To this end, we introduce integer indices i, j, k, which take values $0 \le i \le N_x$, $0 \le j \le N_y$, $0 \le k \le N_z$, such that $\psi(x_i, y_j, z_k)_n \equiv \psi_{i,j,k,n}$, where $x_i \equiv \left(-\frac{N_x}{2} + i\right) h_x$, and similarly for y and z. We also approximate second–order spatial derivatives in the standard way via the central difference formula:

$$\frac{\partial^2 \psi(\mathbf{r})_n}{\partial x^2} \to \frac{1}{h_x^2} \left(\psi_{i+1,j,k,n} - 2\psi_{i,j,k,n} + \psi_{i-1,j,k,n}\right) , \tag{22}$$

with the error of the order of h_x^2. We consider large enough grids such that we can use boundary conditions $\psi(0, j, k)_n \approx 0, \psi(N_x, j, k)_n \approx 0$ and similarly in the other two spatial directions. As a result of using approximation (22) in equation (19), we obtain a tridiagonal system of equations

$$-A\psi_{i+1,j,k,n+\frac{2}{4}} + B\psi_{i,j,k,n+\frac{2}{4}} - A\psi_{i-1,j,k,n+\frac{2}{4}} = \delta_i , \tag{23}$$

for each j and k, where $A = \frac{i\varepsilon}{4h_x^2}$, $B = 1 + \frac{i\varepsilon}{2h_x^2}$ and

$$\delta_i = A\psi_{i+1,j,k,n+\frac{1}{4}} + B^*\psi_{i,j,k,n+\frac{1}{4}} + A\psi_{i-1,j,k,n+\frac{1}{4}} .$$

A solution of this tridiagonal system of equations can be cast in the form

$$\psi_{i+1,j,k,n+\frac{2}{4}} = \alpha_i\psi_{i,j,k,n+\frac{2}{4}} + \beta_i , \tag{24}$$

and from this ansatz we find the recursive relations for the solution:

$$\alpha_{i-1} = \frac{A}{B - A\alpha_i} , \quad \beta_{i-1} = \frac{\delta_i + A\beta_i}{B - A\alpha_i} . \tag{25}$$

From the boundary condition $\psi_{N_x,j,k,n+\frac{2}{4}} = \psi_{N_x,j,k,n+\frac{1}{4}} = 0$, we derive initial values for α and β, $\alpha_{N_x-1} = \beta_{N_x-1} = 0$, which we use to solve the recursive equations (25). With another boundary condition $\psi_{0,j,k,n+\frac{2}{4}} = 0$, we finally solve equation (24). Similar procedure is afterwards followed in y and z direction. Due to the tridiagonal form of the above system of equations, the complexity of the overall algorithm is proportional to the number of discretization points, i.e., $O(N_xN_yN_z)$.

Once the ground–state solution $\psi_0(\mathbf{r})$ is calculated using the imaginary–time version of the above algorithm, the corresponding value of the ground–state energy E_0 of the system can be calculated by numerically integrating expression (5). The described algorithm is implemented using the Fortran programming language [22] and the C programming language [23]. Further computational optimization is achieved by employing the state–of–the-art parallelization techniques for different computing platforms: OpenMP (for shared memory multiprocessing) [25] and hybrid OpenMP/MPI (for distributed memory systems, where each compute node uses OpenMP to maximize performance) [26].

In order to solve the dipolar NLS equation (11) numerically, we need to deal with the integral on the right–hand side of this equation efficiently. To this end, we exploit the fact that the Fourier transform of a convolution integral is given by a product of Fourier transforms of its components

$$I_{\rm dd}(\mathbf{r},t) = \int d\mathbf{r}' V_{\rm dd}(\mathbf{r}-\mathbf{r}')n(\mathbf{r}',t) = \int \frac{d\mathbf{k}}{(2\pi)^3} e^{-i\mathbf{k}\mathbf{r}} \tilde{V}_{\rm dd}(\mathbf{k})\tilde{n}(\mathbf{k},t)\,, \quad (26)$$

where we define the direct and the inverse Fourier transformation as:

$$\tilde{A}(\mathbf{k}) = \int d\mathbf{r} A(\mathbf{r}) e^{i\mathbf{k}\mathbf{r}}, \quad A(\mathbf{r}) = \frac{1}{(2\pi)^3}\int d\mathbf{k}\tilde{A}(\mathbf{k})e^{-i\mathbf{k}\mathbf{r}}. \quad (27)$$

The Fourier transform of the geometric factor (10) stemming from the dipolar interaction can be obtained analytically and is given by

$$\tilde{V}_{\rm dd}(\mathbf{k}) = \frac{4\pi}{3}\left(\frac{3k_z^2}{\mathbf{k}^2} - 1\right). \quad (28)$$

The Fourier transform of $n(\mathbf{r}',t)$ and the inverse Fourier transform from equation (26) are calculated numerically using optimized libraries such as FFTW [27]. Finally, the obtained numerical value $I_{\rm dd}(\mathbf{r},t)$ of the integral (26) is used in the first part of the split–step approach (18) as:

$$\psi(\mathbf{r})_{n+\frac{1}{4}} = \exp\left[-i\varepsilon\left(V(\mathbf{r}) + g|\psi(\mathbf{r})_n|^2 + g_{\rm dd}I_{\rm dd}(\mathbf{r})_n\right)\right]\psi(\mathbf{r})_n\,. \quad (29)$$

Numerical codes implementing the described algorithm for solving the time–dependent dipolar NLS equation are written in the Fortran and in the C programming languages [28]. CUDA parallel version of the algorithm is developed for single GPGPUs [29], as well as a hybrid CUDA/MPI version for distributed memory systems, where each compute node has GPGPU installed [30].

2.3. Parallelization of the Algorithm

Let us note that the tridiagonal systems of equations (23) can be solved separately for different values of j and k. This feature is essential and can be directly exploited to parallelize numerical implementation of the above described algorithm. In fact, this has been done for practically all available modern computer architectures and platforms, and we briefly outline all implementations.

Initial algorithm for solving the NLS equation with the contact interaction term using Crank-Nicolson method was published in reference [22], which also

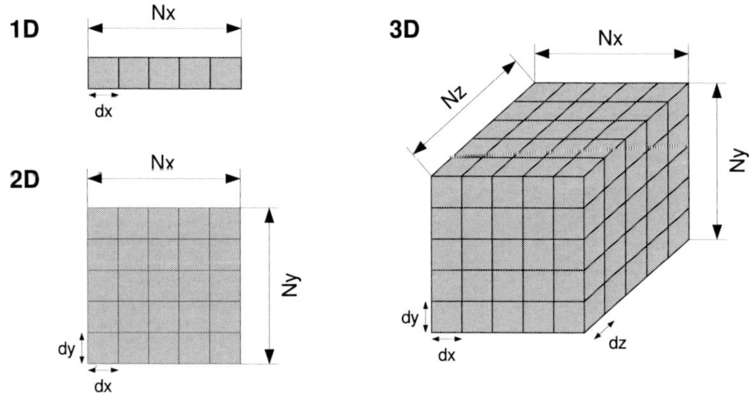

Figure 1. Example of 1D, 2D and 3D meshes with `Nx`, `Ny` and `Nz` discretization points in x, y and z direction, respectively, and the corresponding spacings `dx`, `dy` and `dz`.

includes a serial implementation in Fortran. This was followed by the implementation written in C [23], which includes serial and parallel implementation for shared memory systems consisting of one or more multi-core CPUs using OpenMP approach. Fortran OpenMP implementation was also provided [25], as well as a full parallel MPI version for distributed memory systems, i.e., computer clusters with multi-core CPU nodes [26]. An extension of the algorithm to include the dipolar interaction term was published in reference [28], where the authors provided two serial implementations, in C and Fortran. This was followed by an implementation written in Nvidia CUDA for hardware accelerators in the form of GPUs [29], and later by a comprehensive suite of programs that includes OpenMP, MPI, CUDA, as well as their hybrid combinations [30].

To introduce the parallelization strategy, we start with the description of the discretization scheme used by the algorithm. Let `Nx`, `Ny` and `Nz` be the number of points in each direction, corresponding to the x, y and z directions in one, two or three dimensions (1D, 2D and 3D, respectively). In other words, the number of points in 1D is defined by `Nx`, in 2D by `Nx` and `Ny`, and in 3D by `Nx`, `Ny` and `Nz`. Discretization points in each direction are equidistant, with their spacing defined in dimensionless units by variables `dx`, `dy` and `dz`, illustrated in Figure 1.

The mesh determines how the space is discretized, but the actual values of

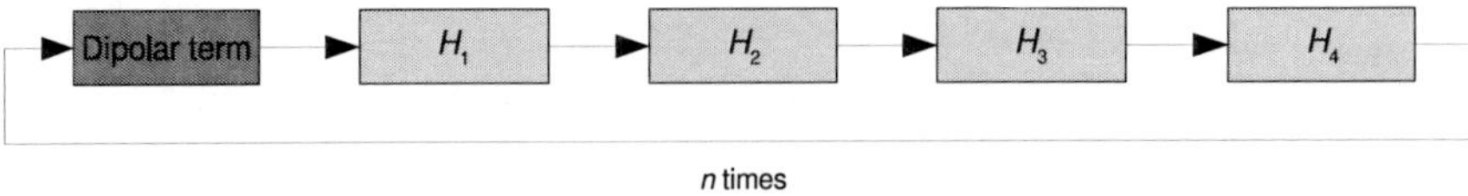

Figure 2. Main loop of the real-time propagation algorithm in 3D.

the wave function, which is also discretized by the mesh, need to be stored in a separate array. For convenience, we refer to this array as `psi` variable. This variable is defined as a vector of size `Nx` in 1D, as `Nx` $\times$ `Ny` matrix in 2D and as `Nx` $\times$ `Ny` $\times$ `Nz` tensor in 3D. In real-time propagation, the wave function values are complex, while in imaginary-time propagation they are purely real, which determines the type of `psi` variable.

Time is discretized according to $t_n = t_0 + n\varepsilon$, with the time step ε stored in a variable `dt`. Taking into account the time step and the total amount of physical time we wish to simulate, we determine the number of iterations required. The main part of the algorithm is a loop where each iteration corresponds to one time step of the propagation. Inside the loop, we update the wave function by propagating it in time with respect to different parts of the Hamiltonian, according to the split-step scheme described in section 2.2 This is illustrated in Figure 2.

In each step of the main loop, wave function values are computed along the discretization mesh in several smaller steps. First such substep is the calculation of the dipolar interaction term (for programs where it is taken into account, otherwise it is omitted). We have two further substeps in 1D, pertaining to propagation w.r.t. H_1 and H_2. In 2D and 3D we follow a similar procedure and add further substeps for H_3 and H_4. Each substep produces intermediate values of the wave function that are used in the next substep. The order of the substeps dealing with parts of the Hamiltonian with spatial derivatives (H_2, H_3 and H_4) can be arbitrary, a feature useful in hybrid and distributed memory algorithms. Propagation of the wave function w.r.t. H_1 relies on the availability of the dipolar interaction term value, which is computed in the first substep, before we start with the update of the wave function.

When necessary, the dipolar interaction term is computed using the discrete Fourier transform (DFT). DFT of a sequence is commonly implemented by relying on the *Fast Fourier transform* (FFT) algorithm via an external library. From equation (26) it follows that we first need a DFT of a sequence consisting

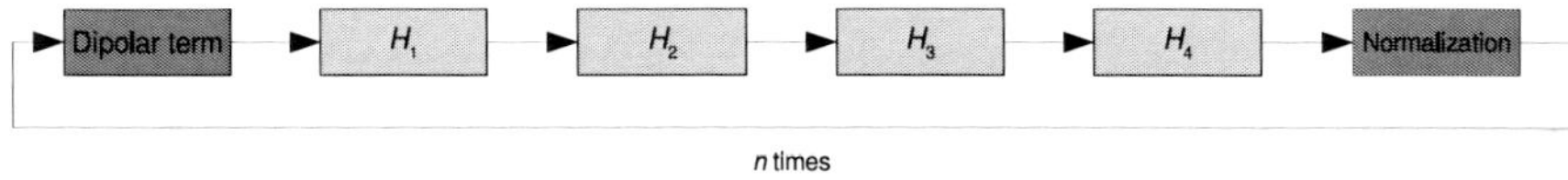

Figure 3. Main loop of the imaginary-time propagation algorithm in 3D, which introduces an additional step of normalization.

of absolute squares of wave function values. The resulting complex-valued sequence is then multiplied by the Fourier transform of the dipolar potential. This transform is a known function (28), precomputed for a chosen spatial mesh and stored in an array. Next, by performing the inverse DFT we get the dipolar interaction term stored in the resulting array, which we then use to propagate the wave function w.r.t. H_1. In 2D and 3D, we need multidimensional DFT, which can be computed by the composition of a sequence of 1D DFTs along each direction.

The substep in which the wave function is propagated w.r.t. the Hamiltonian part without spatial derivatives (H_1) proceeds by employing further nested loops over mesh points. Part of the H_1 is the trap potential $V(\mathbf{r})$, which, like Fourier transform of the dipolar potential, can be computed only once during the initialization, and then reused. We assume that the trap potential is static, but in case of a time-varying trap potential, the algorithm can be modified to update the trap potential before this substep, i.e., at the beginning of each iteration. The remaining substep, or substeps in 2D and 3D, propagates the wave function by relying on the Crank-Nicolson scheme. We need to perform a backward sweep of the mesh in each direction to determine the corresponding Crank-Nicolson coefficients α, β and δ for each direction, followed by a forward sweep to determine the solution for the entire space range. The algorithm for imaginary-time propagation is similar to the one discussed above, with the addition of the normalization step at the end of each iteration of the main loop, as illustrated in Figure 3. To normalize the wave function, we first need to compute its norm and then to divide the wave function values with that norm.

We see that the algorithm is computationally very demanding, as it requires multiple passes over the entire mesh in each iteration of the main loop. In 2D and 3D, the total number of discretized points sharply increases, making this problem even greater. As we have identified above, several variables can be moved out of the main loop and computed only once, however the remaining calculations are still very demanding, hence the need for parallelization.

The described substeps cannot be executed concurrently, as the intermediate values they produce are used in subsequent substeps. Instead, parallelization of the algorithm can only be achieved by focusing on the mesh loops inside substeps. This is an obvious choice for shared memory systems, as all data stored in the memory is local and accessible to all participating processes, removing the need for data transfers between processes. The loops themselves can easily be parallelized if they do not contain recursive relations. This is true for the substeps dealing with the computation of the dipolar term and propagation w.r.t. H_1. Substep involving H_2 cannot be easily parallelized in 1D, as it has recursive relations in both backward and forward sweeps. In general, recursive relations can be parallelized using the *scan* algorithm [31] (also known as the generalization of *prefix sum* algorithm [32]), however the implementation complexity of such an algorithm made us discard this approach. In 2D and 3D, we can exploit the fact that the recursive relations appear only in the innermost loops of the substeps involving H_2, H_3 and H_4. Thus, we can achieve parallelization by dividing the work among the processes at the level of the outermost loop.

In addition to the above outlined parallelization that can be naturally implemented on a single computer using OpenMP approach, one can also utilize modern GPUs to parallelize the algorithm on a single node, and even combine CPU and GPU into a hybrid algorithm. We first discuss CUDA-based parallelization, introduced in reference [29]. The significant differences in architecture of the GPU and its accompanying execution model on one side, and CPU architecture with its conventional execution model on the other side, do not result in significant changes to the main algorithm for the case of Nvidia CUDA implementation. From the algorithm's point of view, both CPU and GPU are shared memory systems with multiple processing units, even though they are vastly different, and therefore can be used in a similar manner. Main loop of the algorithm remains unchanged, as well as each substep. However, one has to replace the FFTW library used by C/OpenMP programs with the cuFFT library, which is available in CUDA. cuFFT provides two interfaces: the native one and the FFTW one [33]. FFTW interface is intended to be used as a drop-in replacement for FFTW, allowing programs written primarily with FFTW in mind to use CUDA GPUs with minimal modifications to the source code. While this interface could be used as a temporary solution during the development of CUDA programs, the goal of using CUDA for all computation relating to the propagation of the wave function could only be reached with the native cuFFT interface. This interface is modeled after FFTW but differs from it in the way plans are created and executed. In cuFFT, precision and data type of the transform is determined by the functions initiating execution of plans, which results in a slightly different sequence of function calls and in improved performance.

In order to use CPU and GPU at the same time for hybrid algorithms, the work has to be divided efficiently between them. If we put aside the obvious hardware architecture difference between CPU and GPU, we can consider the computer with GPU as being similar to the distributed memory system consisting of two computing nodes with different characteristics. Both CPU and GPU have their own memory, and are connected through a fast interconnect, in this case the PCI-Express. This means that we need an approach similar to the one distributed memory algorithms use to achieve the desired parallelization. The most important difference here is that all data is available in main (CPU) memory, and only portions of it need to be transferred to GPU memory, as opposed to the true distributed memory systems, where no single part of the system has all the data. Therefore, the general approach to perform this type of hybrid computation is to:

1. designate a portion of data to be processed by GPU,

2. copy that data to GPU while simultaneously using CPU for computation over the remaining data,

3. transfer the data back from GPU, overwriting old data, and

4. synchronize CPU and GPU.

This flow is illustrated in Figure 4.

Note that data can be offloaded from CPU to GPU in various ways [34, 35] and, in general, any data distribution scheme which targets distributed memory systems may be used. However, we settled on a simple approach using 1D decomposition, also known as *slab decomposition*. When using slab decomposition, data is distributed along one dimension, usually the slowest-changing one, and the remaining dimension (in 2D programs) or dimensions (in 3D programs) of data remain local. This means that we can perform computation on the local data efficiently, and no data exchanges are necessary. Depending on a data access pattern, in our programs we rely on decompositions either along the x direction (discretized with the `Nx` spatial points), or along the y direction (discretized with the `Ny` spatial points). In 3D programs, there is no need

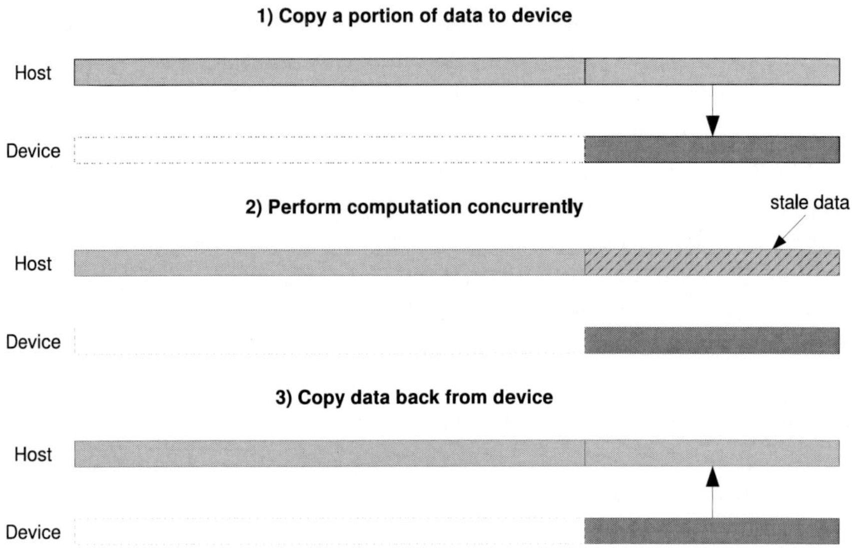

Figure 4. Flow of data between host and device.

to decompose the data along the z direction. In case we need data from the distributed dimension to become local in order to perform some computation (e.g., to update the wave function values when data are decomposed along the x direction), we have to reassemble data in host memory and decompose again along the appropriate dimension. Figure 5 illustrates this concept.

With the decomposition scheme in place, we can consider how to divide the computation of the wave function propagation in time between CPU and GPU. Once the data have been distributed between CPU and GPU, on the CPU side only the exit condition of the outer loop needs to be adjusted, so that the CPU processes a smaller number of elements (e.g., by replacing `Nx` with `cpuNx`). Similar changes are required on the GPU side as well, however we also need to ensure that GPU has valid data to work with, and that data are reassembled in host memory after the computation on GPU is done.

Therefore, in the hybrid implementation we initially distribute the data along the slowest-changing dimension. This allows CPU and GPU to simultaneously perform computation of wave function propagation w.r.t. H_1 and H_3 parts of the Hamiltonian (and H_4 when working in 3D). The propagation of the wave function w.r.t. H_2, corresponding to the x direction, can be done locally

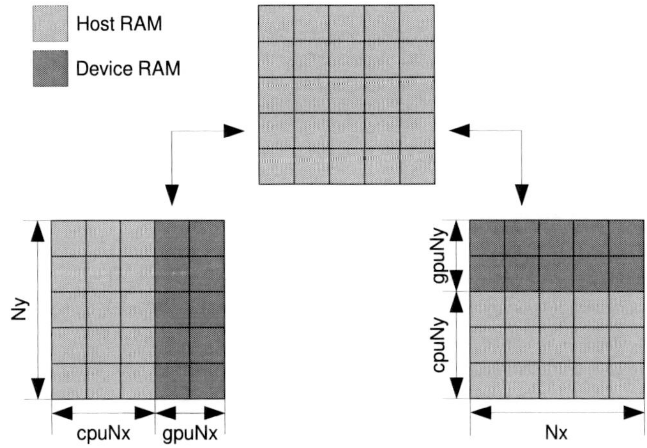

Figure 5. Two offload patterns used. If offloaded along x direction, the y direction remains local. If we need to access whole x direction data, we need to decompose along y direction.

only if we decompose the data along the y direction beforehand. Note that we do not have to transfer data from GPU back to host memory at the end of each function, due to the fact that time propagation w.r.t. H_2, H_3 and H_4 parts of the Hamiltonian can be done in arbitrary order in each step. Without this, the time propagation workflow of 3D programs would have to involve the following steps, according to Figure 2:

1. calculate the dipolar term,
2. transfer the data decomposed along x direction to GPU,
3. propagate the wave function w.r.t. H_1
4. reassemble the data in host memory,
5. transfer the data decomposed along y direction to GPU,
6. propagate the wave function w.r.t. H_2,
7. reassemble the data in host memory,
8. transfer the data decomposed along x direction to GPU,

9. propagate the wave function w.r.t. H_3,
10. propagate the wave function w.r.t. H_4,
11. reassemble the data in host memory.

Rearranging the order of time-propagation substeps allows us to remove one transfer of data to GPU and its subsequent reassembly in host memory:

1. calculate the dipolar term,
2. transfer the data decomposed along x direction to GPU,
3. propagate the wave function w.r.t. H_1,
4. propagate the wave function w.r.t. H_3,
5. propagate the wave function w.r.t. H_4,
6. reassemble the data in host memory,
7. transfer the data decomposed along y direction to GPU,
8. propagate the wave function w.r.t. H_2,
9. reassemble the data in host memory.

Therefore, we use the above, optimized sequence in all 3D programs, and similarly in 2D.

To complete the hybrid algorithm, we also need to distribute computation of the dipolar term between CPU and GPU, where complexity arises in performing DFT on distributed data. Currently available FFT libraries target either CPU or GPU for their computation, but unfortunately not both at the same time. There are numerous attempts to develop specialized FFT libraries which would enable this [36, 37, 38], however a full-featured library with support for R2C transforms with advanced data layout is still not available. The approach we used is to rely on existing libraries for actual transforms, but perform data distributions manually. The libraries we use, FFTW on CPU and cuFFT on GPU, support advanced data layouts as well as working on a subset of the whole data, allowing for an efficient implementation.

To perform the Fourier transform simultaneously on CPU host and GPU device, we have to split the single multidimensional transform into a series of

1D transforms along each dimension of the input data, which can be computed on CPU and GPU independently. This approach is known as the *row-column* algorithm [39], and is often used in FFT libraries. The essence of this algorithm can best be summarized in an example of a 2D FFT. Given a matrix `Nx` × `Ny`, we compute the DFT in the following way:

1. Transfer portion of the input array, decomposed along the x direction, to device memory. We transfer last `gpuNx` × `Ny` consecutive array elements to GPU. This can be done as a single memory copy operation due to the flat allocation that was used. The choice whether to offload data to GPU memory from the beginning or the end of the input array is arbitrary.

2. Perform DFT along the y direction on both CPU and GPU concurrently. CPU will perform `cpuNx` such transformations, while the GPU will perform `gpuNx` 1D DFTs. Each of these transforms will take a subset of the input array, which are `Ny/2 + 1` elements apart. Note that CPU will not do anything with the last `gpuNx` × `Ny` elements.

3. Copy the array with the transform back from GPU to CPU, writing over the stale data with the relevant portion transformed on GPU. After this step, we have the complete array transformed along the y direction residing in host memory.

4. Transfer portion of the input array, decomposed along the y direction, to GPU memory. Similarly to step 1, we transfer last `Nx` × `gpuNy` elements to GPU.

5. Perform DFT along the x direction on both CPU and GPU. This time, CPU will perform `cpuNy/2 + 1` such transformations, while GPU will perform remaining `gpuNy/2` transformations. The halving of the number of transformations is due to use of R2C transformations. In each transform, elements are `Ny/2 + 1` places apart, while the first element of each transform is adjacent to the previous one.

6. Copy the array with the transform back from GPU to CPU. With this step completed, we have the full FFT of the input array residing in host memory. This step may be omitted if the computation that follows does not require the FFT of input data to be fully assembled in host memory. This is the case in our algorithm, as we use the resulting transformed array

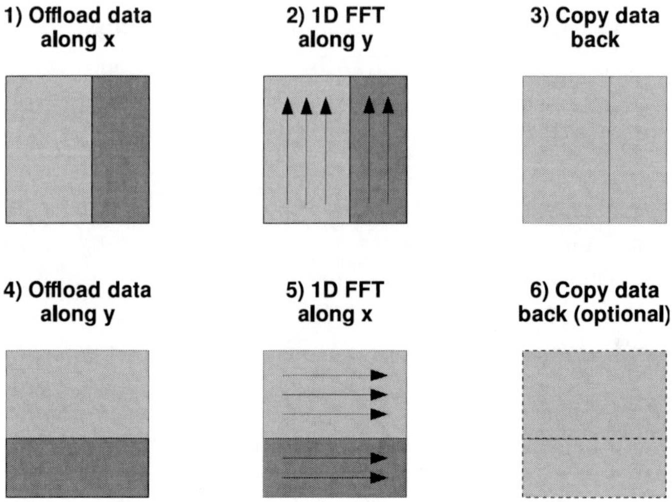

Figure 6. Hybrid algorithm for concurrent FFT on host and device.

only for subsequent computation on GPU. Since the memory transfer is expensive, omitting this step leads to significant performance improvement.

The inverse Fourier transform can be done in an analogous way. In 3D, the general principle remains the same, however, we do not separate the DFT into three 1D transforms, but to one 2D and one 1D transform. We do this due to better performance of 2D transforms, which in both libraries used is found to be better than two 1D transforms. Figure 6 illustrates the DFT algorithm described above.

And finally let us mention that several distributed memory implementations, targeting computer clusters, were developed as well:

1. pure CPU version, which is built on top of the shared memory algorithm (OpenMP/MPI),

2. pure GPU version, which is built on top of the CUDA implementation of the shared memory algorithm (CUDA/MPI),

3. hybrid version, which is based on the algorithm described in Chapter (Hybrid/MPI).

Two of the three implementations, the OpenMP/MPI and CUDA/MPI implementations, have been published in reference [30] and are available for download [40, 41]. The Hybrid/MPI implementation is also publicly available [41]. The main challenges in the development of distributed memory and hybrid algorithms were devising an efficient data distribution scheme that facilitates fast data shuffling between computing resources. In the MPI implementations this was done via a transpose operation realized in two ways, one relying on FFTW and the other relying on MPI vector types and collective communications. Further details can be found in reference [42].

2.4. Performance of Parallel Algorithms: Measurement Results and Modeling

In previous section we presented various parallel implementations of programs solving the dipolar and contact interaction GP equation without providing details of their measured performance. This section is dedicated to filling that gap. We tested various scaling scenarios, strong and weak, from a single computing node to the cluster. All testing was done at the PARADOX supercomputing facility located at the Scientific Computing Laboratory, Center for the Study of Complex Systems of the Institute of Physics Belgrade.

Performance of hybrid implementations is highly dependent on the amount of work offloaded to GPU(s), so the key to maximizing performance of these versions is proper selection of parameters which control this feature. We used evolutionary computation techniques to find the optimal solutions, as described here. In the remaining subsections we give an overview of the tests and the methodology they relied on, present the performance results and their modeling, and finally discuss how to select the optimal algorithm for a given hardware platform.

2.4.1. Optimization of Input Parameters for Hybrid Implementations

Hybrid implementations, running on a single computer or a cluster with multiple computing nodes, can potentially offer the best performance of all algorithms presented in this thesis, by utilizing all allocated computing resources. For this to happen, work must be divided between CPU and GPU in such a way as to maximize their throughput and minimize their idle time. A question that naturally arises is how to achieve this for any possible mesh size for a given CPU/GPU combination. Unfortunately, there is no single best way to divide the

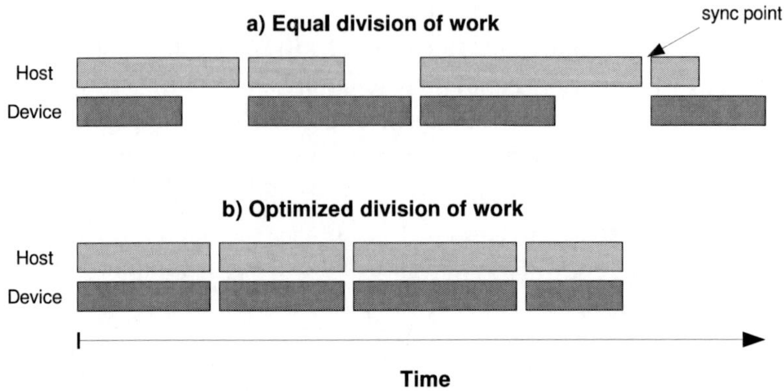

Figure 7. Execution timeline in a hybrid system: (a) with equal data distribution, leading to unbalanced computation time; (b) with optimized data distribution and ideal computation load.

work, since processing powers of CPUs and GPUs vary significantly. For example, in a computer where a powerful new GPU is paired with an older type of CPU, more work should be offloaded to the GPU, and vice versa for computers with a powerful CPU and low-performing GPU. Even if the installed CPU and GPU offer similar performance in terms of floating point operations per second (as is the case for the PARADOX cluster), the differing architectures mean that certain portions of the algorithm are better suited to CPU or GPU. If we were to divide the work naively, e.g., equally, this would lead to unbalanced computation, as illustrated in Figure 7(a), which in turn increases the computation time.

To get over this problem, the described hybrid implementations have an option to manually specify each parameter that controls the work being done on a GPU. Ideally, this flexibility would allow us to overlap computation on CPU and GPU as much as possible, thus minimizing the execution and idle time of each resource. The execution timeline for the ideal case looks like the illustration in Figure 7(b).

The aforementioned parameters include the total amount of data transferred to GPU, the number of chunks, kernel grid size parameters and a special parameter controlling whether some functions will even be offloaded to GPU or not, which is useful in situations where the CPU is much more powerful than the

GPU. In 3D programs, this amounts to 33 integer-valued parameters for which we would like to find the optimal values for the desired mesh size on a given computing system (with given hardware characteristics). Furthermore, there are constraints that the parameters must satisfy. For example, the total amount of data transferred to GPU must be divisible by the number of chunks, and the size of a block that kernel launches must fit within the limits imposed by the underlying GPU.

To evaluate any combination of parameters, the programs have to be executed, and their execution time measured. The search space of 33 parameters is clearly too large to be exhaustively traversed in a reasonable amount of time. Fortunately, we can reduce the number of parameters we have to search through at any single time by grouping them based on the synchronization point. After the CPU and GPU synchronize, data have to be divided again, so the only relevant parameters in the region between two synchronization points are the ones controlling the division of work and the ones controlling the kernels which are executed in that region. During a single iteration, CPU and GPU have to synchronize four times:

1. after the FFT on local dimensions,
2. after the inverse FFT on the non-local dimension,
3. before the function performing propagation w.r.t. H_2, and
4. after that function (which represents the end of a single iteration).

This gives us four distinct parameter sets to optimize: PS_1 with 6 parameters, PS_2 also with 6 parameters, PS_3 with 15 parameters and PS_4 with 5 parameters. We can consider each parameter set independently.

We investigated several approaches to optimization of parameter sets: naive brute-force search, iterative optimization via gradient descent, and metaheuristic via genetic algorithm. Our goal was not to find the best possible optimization algorithm for our problem, but rather to implement a reasonably good one. By this we mean the algorithm which is not difficult to implement, which will give us a set of parameters that are close to the optimal ones, and will not take a long time to finish. Our focus was on 3D variants of both single node and MPI versions of hybrid programs. The concepts presented here apply equally to 2D variants and there is no important difference between the single node and the MPI version, so we will not make this distinction in the remainder of this section. We will now describe how the three approaches were implemented.

A brute-force search (BFS), or exhaustive search, considers every possible combination of parameter values in order to find the optimal one. Even with parameters divided into smaller sets, the exhaustive search takes too long, and is only feasible if we narrow down the ranges of all parameters involved. Unfortunately, this is only possible when we have a deep understanding of the performance of the CPU and GPU models used, when we can provide a reasonable guess for the optimal solution. An alternative is to consider only certain values along the range of a single parameter, rather than every value in the range. This approach is feasible and we were able to find reasonably good parameter values using it. The downside is that this way we may miss the optimal values of parameters, as they often lie between the selected test points. For instance, on one PARADOX computing node and a mesh size $256 \times 256 \times 256$, with the work divided between CPU and GPU along the outermost, x dimension, if we test only for the values of offloaded data to GPU that are multiples of 32 (i.e., $32 \times 256 \times 256$, $64 \times 256 \times 256$, etc.), we will miss the optimal value which lies around $140 \times 256 \times 256$ (assuming all other parameters are fixed). Even though BFS is often not feasible, it is useful as the baseline for the evaluation of other methods we implemented.

With the BFS implemented as a baseline, we switched our focus to the implementation of an iterative optimization algorithm. A well-known and widely used algorithm is *gradient descent* (GD) [15]. It is an iterative procedure in which every iteration aims to get closer to the minimum of the given function $F(\mathbf{x})$, where $\mathbf{x}$ represents a vector of function parameters. If the function $F(\mathbf{x})$ is defined and differentiable around some point $\mathbf{a}$, then $F(\mathbf{x})$ decreases fastest in the direction of the negative gradient of F at $\mathbf{a}$. GD exploits this fact and defines $\mathbf{b}$ as

$$\mathbf{b} = \mathbf{a} - \gamma \nabla F(\mathbf{a}) \,, \tag{30}$$

so that $F(\mathbf{b}) \leq F(\mathbf{a})$, for γ small enough. This observation is used as a building block for the algorithm. We start from an initial guess $\mathbf{x}_0$ for a location of the minimum of F and construct a sequence $\mathbf{x}_n (n \geq 1)$ such that:

$$\mathbf{x}_n = \mathbf{x}_{n-1} - \gamma \nabla F(\mathbf{x}_{n-1}) \,. \tag{31}$$

This sequence will eventually converge to the local minimum, if one exists. GD is most applicable to functions for which the gradient can be computed analytically, however it can also be used when the derivative can only be obtained numerically. To apply it here, the execution time of our programs will be the function F that we minimize, and its arguments are the $n = 33$ parameters that control the GPU. The execution time is measured as the average time of a single iteration of the main time-propagation loop, sampled over a given number of iterations. As explained earlier, we can divide the problem into four separate minimization problems, with four sets of parameters of dimensionality m_i ($i = 1, 2, 3, 4$), in the same way as in BFS. To numerically compute the k-th partial derivative of one of the minimization functions at a set of candidate parameters $a_1, a_2, \ldots, a_m$, we use first order approximation:

$$\frac{\partial F}{\partial x_k}(a_1, \ldots, a_m) \approx \frac{F(a_1, \ldots, a_k + h_k, \ldots, a_m) - F(a_1, \ldots, a_k, \ldots, a_m)}{h_k}, \tag{32}$$

where h_k is the increment of a given parameter. Since the parameters are integer-valued, we have to carefully choose the value of h_k, to be as small as possible, while still satisfying all the constraints that the corresponding parameter may have w.r.t. other parameters. We note that in order to evaluate the full gradient of F at $(a_1, \ldots, a_m)$ we need to execute our programs $m + 1$ times.

We stress that execution times of our programs are not always the same due to the hardware, software and OS scheduling issues, making the minimization functions noisy. This noise affects the calculation of the gradient and could point the algorithm in the wrong direction. We can detect this by checking if the value of minimization function has increased in subsequent GD iteration, and discard this move if necessary. While useful, resorting to this tactic means that in case the GD gets stuck in a local minimum, it will not be able to get out and thus will never reach the global minimum. A naive way to check if this is the case would be to start over from a different initial point, and see if the GD algorithm converges to the same minimum. There are more sophisticated approaches to addressing this issue, which we discuss later.

In general, it is not possible to completely eliminate noise, due to inherent problems of accurately measuring time on a computer. However, we can try to minimize it by increasing the precision of execution time measurement of our programs. This can be achieved by averaging execution times over larger number of iterations of the main time-propagation loop. Unfortunately, this also means that the total execution time of the GD algorithm will significantly increase.

Another issue in our implementation of GD arises when we have to select the next values of our parameters based on the output values of previous GD iteration. These output values are real-valued, meaning that we need to convert them to integers somehow. Simply rounding them up or down to the nearest integer will not be enough, especially for small value of γ. For example, if we set the initial value of some parameter to 10, and after one GD iteration the proposed value of that parameter is 10.2, rounding it down will reset the value back to 10, effectively discarding the whole GD iteration. To avoid getting stuck in this way, our implementation selects the next possible value in the direction of change. In the example above it would mean selecting 11 as the new value. This change makes it impossible for the algorithm to converge, but it will usually stay close the minimum where all proposed solutions are of similar quality.

With the GD algorithm implemented as described above, we are able to get a set of optimized parameters from a random set of initial values. However, such optimized parameters are often suboptimal, and do not have the best performance. This is due to the fact that in most cases the GD will converge to the nearest local minimum, and there is no guarantee that that local minimum is also the global one. Using our previous example from BFS, a mesh of $256 \times 256 \times 256$, with GD we obtain that local minima exist for all power of two values, i.e., parameters suggest offloading $2^n \times 256 \times 256$ to the GPU. If we randomly select a small initial value, GD will get stuck in the nearest local minimum, which in this example will be far from the global one, thus producing a very bad solution. We can attempt to avoid getting stuck in a local minimum by adapting the parameter γ, also known as *learning rate* in machine learning literature. Keeping the learning rate high for a first few iterations would allow the algorithm to find the general location of the minimum, after which we could gradually lower the learning rate until convergence is achieved, in a process called *annealing* [43]. In practice this did not completely solve the problem, as we found that the optimal learning rate to start with varies with the initial set of parameters and the mesh size, and thus has to be manually selected. This complicates attempts to automate the process of finding the optimal parameters for a range of mesh sizes, which we needed as part of the tests of both hybrid implementations. A method to remove the manual tuning of the learning rate exists [44], but we find it to be too complex to implement for our programs, because it would require significant changes to the way the parameters are selected.

From the discussion above, we conclude that the GD is not the best-suited method for the optimization in our case, mostly because our minimization function is noisy and the numerical computation of the gradient is costly. Derivative-free optimization methods would be better suited to the problem, e.g., stochastic approximation algorithms like simultaneous perturbation (SPSA) [45, 46], or metaheuristics like genetic algorithms. We decided to use a genetic algorithm approach, as it is simple to understand and implement, as well as easy to adjust to get the desired behavior.

Genetic algorithm (GA) is an optimization method based on natural selection that mimics the process found in biological evolution [47]. GA works by creating a population of individual solutions, which it then evaluates and modifies, creating a new population, and iterating this process. Unlike the classical algorithms like GD, which iterate a single candidate solution towards the optimal one, the GA iterates a population of solutions in which the best individuals approach the optimal solution. The initial population is usually created at random, giving the GA different points in the search space to start from. The individual solutions are evaluated using a user-supplied fitness function, giving each individual a score based on how well they perform the given task. Individuals with the highest score are then selected to "reproduce" and create new offspring, after which they may be mutated randomly. The offspring form a new population, and the process can be repeated again. The GA continues until a suitable solution is found, or after a certain number of generations has passed.

To implement a GA, we first have to decide how to represent an individual. Individuals are created based on their blueprint, called *chromosome* in GA terminology. The most often used representation is bit-string [48], where all properties of an individual (its *genes*) are serialized to an array of bits, and then concatenated. More advanced representations exist, e.g., for encoding real values, permutations and general data structures [49, 50], however since our individuals are sets of integer parameters, we did not develop any special representation and instead we used ordinary arrays of integers as a chromosome. Therefore, each gene in a chromosome is a single GPU parameter of our programs. Individuals need to be evaluated using a fitness function. In our case, this means executing the programs with the parameters extracted from the individual's chromosome, and reporting the execution time as fitness score, with lower execution time being better.

Next step is the implementation of the three GA operators: selection, reproduction (*crossover* in GA terminology) and mutation. Each operator can be implemented in different ways. Most common type of the selection operator is *roulette wheel* selection. This is a type of fitness-proportionate selection, with the idea to give each individual a slice of the circular roulette wheel based on their fitness. The wheel is then spun, and when the roulette ball stops, the individual in whose region the ball stopped is selected. In this way, the fittest individuals have the greatest chance of being selected for reproduction, while the ones with very low score quickly die out. We also tested an alternative type of selection, the *tournament* selection. In this type of selection, we randomly select several individuals, and host a tournament for them. The individual with the best fitness score wins, and is selected for reproduction. By changing the size of the tournament we can control the selection pressure, e.g., by using small tournament size in the first few iterations we can prevent premature convergence and increase it in later generations when we have explored the search space enough. In our tests, the tournament selection gave slightly better results for smaller populations, by keeping the population diverse in early generations. For larger population size, both selection rules performed equally. Alongside the main selection algorithm, we also used *elitist* selection, where first few individuals are copied to the next generation without changing their chromosome. This prevented the loss of the best individuals in the next generation, but has to be used carefully as it may lead to premature convergence to a suboptimal solution.

To produce the next generation, selected individuals should combine their chromosomes and produce offspring. This is the task of the crossover operator. Crossover exchanges parts of the chromosomes, mimicking the biological recombination from nature. In its most basic form, crossover works by randomly selecting a point and exchanging segments before and after the point to create two new offspring from two parents. This type of crossover is called *single-point crossover*, and is illustrated in Figure 8(a). Other popular crossover techniques are *two-point crossover* and *uniform crossover*, illustrated in Figures 8(b) and 8(c). Two-point crossover is similar to the single-point crossover, just with two points instead of one. On the other hand, in the uniform crossover each gene of the offspring is selected randomly, either from the first parent or from the second one, with some fixed probability, typically 0.5. Using the uniform crossover leads to a wider exploration of the search space [51], but this may not always result in better performance of the operator [52]. We tested single- and two-point crossover and found that there is little difference in terms of performance between them. Since some of our parameters have constraints, it is important to select only the crossover points which lead to an allowed recombination of genes. Unfortunately, this made the implementation of uniform crossover impossible, and we did not pursue it further.

The final step in producing the next generation is to apply the mutation operator on the new population. The mutation operator randomly changes genes

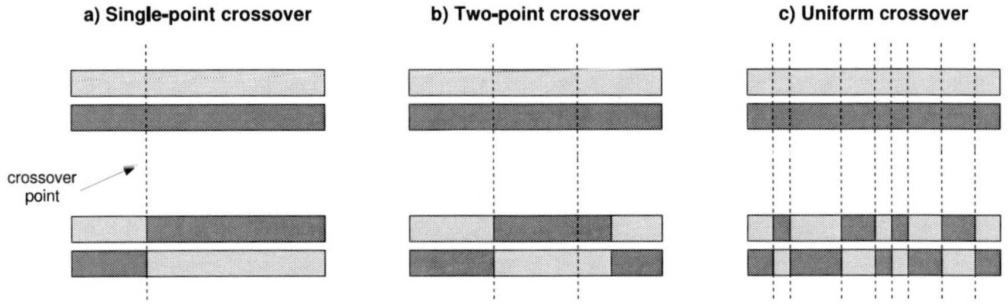

Figure 8. Three different crossover techniques in GA: a) single-point crossover, b) two-point crossover and c) uniform crossover.

to new values, which is equivalent of performing a random walk through the search space. Changing every gene would not be desired, so mutation operates with a very small probability, meaning most of the genes will be left unchanged. When chromosomes are implemented as bits, the mutation would be equivalent to flipping a random bit. However, we could not implement mutation this way, as random changes to the chromosome would often result in non-functioning individuals, e.g., parameter constraints would not be satisfied. Special care must be taken to ensure that the mutation produces a healthy individual, similarly to the crossover operator. The importance of mutation operator and its relation to the crossover is often debated [53, 54], with its role being defined as "to maintain diversity within the population and inhibit premature convergence" [55], as the crossover operator does not introduce new information to the population. In our tests, the mutation plays a crucial role if the population size is small, e.g., less than 20 individuals, in accordance with references [56, 57]. If the population size is large, e.g., over 200 individuals, the positive effects of mutation on the population fitness are not that evident.

GA is typically iterated for a fixed number of generations, as is the case in our implementation. Alternatively, we could have implemented some exit clause in the main loop of the GA which would stop the evolution after the best individuals have not been improved for some number of generations. After each run of the GA, there is usually several highly fit individuals in the population. Since randomness is an integral part of any GA, different runs of the algorithm produce slightly different results. Therefore, our GA does not converge to a single solution, but produces candidate solutions that have very similar parameter values and overall fitness. Among the candidate solutions, there may be small differences in the amount of data offloaded to GPU or the kernel parameters, with the execution times negligibly different. The number of generations required to produce a good solution also varies due to the randomness in the initial population and the population size. In our tests, it has a value between 10 and 50.

The implementation described above provides only the basis for a successful application of GA to the problem at hand. As can be seen, there are several important parameters of GA to tune in order to get optimal results. These include the population size, the number of generations, mutation rate, and the selection parameters (e.g., elite selection rate, the tournament size). We did not perform thorough testing of the performance of the GA that would allow us to obtain the best values for these parameters, as we were more focused on the quality of the candidate solutions GA creates. However, we observed that our GA finds good solutions faster if the initial population size is between 100 and 200, with the number of generations between 10 and 20, depending on the set of parameters we wish to optimize.

To get a perspective of how the three optimization methods perform, we tested them by comparing their final solutions, as well as by recording the number of program executions needed to get to the optimal set of parameters. For this test, we used 3D real-time propagation hybrid program on a single computing node. All three methods were allowed to execute the program up to 1000 times. The mesh sizes used range from $80 \times 80 \times 80$ to $600 \times 600 \times 600$, all of which could potentially be offloaded to GPU. We tested parameter set PS_3 (with 15 parameters), which controls the execution of FFT and the kernels in the subsequent functions that perform propagation w.r.t. H_1, H_3, and H_4. The BFS algorithm was used as the baseline. Note that the range of parameters and a small number of allowed executions implies that the BFS algorithm takes the values of parameters with large stride, potentially missing the optimal solution. GD method is used as described, with the learning rate γ initially set to a higher value, which was gradually decreased. We performed the GD for 60 iterations, which amounts to 960 program executions. The GA was run on a population of 100 individuals, for 10 generations, amounting to the same the number of program executions as the BFS. We have used mutation rate of 5%, and the tournament selection, with the tournament size equal to 10% of the population size. Elitism was also included, with the top 2% of the population copied over. All algorithms were tested five times, to minimize the effects of random initialization of both GD and GA methods. The results are shown in Table 1.

Table 1. A comparison of three optimization methods. Reported times are given in milliseconds, for a single iteration of the main time-propagation loop, averaged over 50 iterations. The last column contains the minimal execution times obtained by manual tuning. The reported time for each algorithm represents the minimal achieved value in five test runs

Mesh size	BFS	GD	GA	Best
$80 \times 80 \times 80$	7	6	6	6
$128 \times 128 \times 128$	18	27	16	16
$240 \times 240 \times 240$	154	272	135	126
$256 \times 256 \times 256$	181	199	169	156
$360 \times 360 \times 360$	343	387	312	298
$480 \times 480 \times 480$	981	1049	868	829
$512 \times 512 \times 512$	1452	1628	1312	1242
$600 \times 600 \times 600$	2591	2984	2227	2159

From the results we conclude that the GA was the most effective optimization method of the three approaches. As expected, GA found better solutions than BFS due to large strides BFS had to use. The randomness in the initial population has a big effect on the convergence of the GA method, sometimes enabling it to find the optimal solution after just three generations. Even if the fitness of the initial population is very bad, the GA still converges to very good solutions after 10 generations. On the other hand, GD performed very poorly, mainly because the noise in program execution times has often thrown it in the wrong direction. Also, GD would get stuck in the nearest local minimum, which often was not the global one. When the initial position of GD is near the global minimum it converged to toward the optimal solution, which was rarely the case. For mesh sizes larger than the tested ones (relevant for MPI-based implementations), corresponding to a larger range of parameters, GD would be even less effective. Both BFS and GD can be made more usable if the range of parameters can be narrowed, i.e., if we know the relative performance of the GPU in comparison to the CPU. However, the GD would still be somewhat inefficient, due to its higher susceptibility to noise.

Since the GA method is shown to be superior to the other two approaches, we use GA as our optimization method of choice in the next two sections.

2.4.2. Testing Methodology

All programs were tested on the PARADOX-IV cluster, which is a part of the PARADOX supercomputing facility. This cluster is comprised of computing nodes with two Intel Xeon E5-2670 Sandy Bridge CPUs (with a total of $2 \times 8 = 16$ cores), with 32 GB of RAM and one Nvidia Tesla M2090 GPU with 6 GB of GPU RAM, each connected by InfiniBand QDR interconnect. We used Intel's compiler (version 2016) to compile the serial and OpenMP programs, and CUDA 7.5 for the GPU portions of the CUDA and hybrid programs. MPI-based implementations were compiled with Open MPI (version 1.10), which itself relied on underlying Intel and CUDA compilers. In the case of Hybrid/MPI programs, we performed tests using both FFTW and our own transpose routines. We found the minimal execution times to be about the same for both approaches, but the FFTW transpose would sometimes exhibit very bad performance due to the creation of suboptimal communication plan. For this reason, the execution times reported for Hybrid/MPI implementation are obtained using only our own transpose routines.

The base of all performance evaluations was the measured execution time of critical regions of the programs, i.e., the portions performing wave function propagation in imaginary or in real time. This measurement excluded the time spent in other parts of the programs, e.g., initialization of OpenMP/CUDA/MPI environment, memory allocation and deallocation, creation and destruction of FFTW plans, initialization of variables and I/O operations. Measuring average execution time of a single iteration of the main time-propagation loop allows us to predict the performance and total execution time of a given simulation, as the number of iterations is specific to the problem at hand and may vary significantly between different simulations. All measurements were collected using high precision timers based on `clock_gettime` POSIX function on the CPU side, and CUDA event API on the GPU side. The execution time of a single iteration of serial programs depends on the mesh size, controlled by variables `Nx`, `Ny` and `Nz`. For parallel programs, we can measure speedup and scaling efficiency as a function of the varying number of processing elements (OpenMP threads or MPI processes).

We tested performance of 1D, 2D and 3D programs on a range of mesh sizes, for a varying number of OpenMP threads and MPI processes, as shown in Table 2. Mesh sizes were chosen from the corresponding range, and we did not focus solely on mesh sizes which maximize performance of the programs, to obtain a more realistic assessment. We varied the number of OpenMP threads from 1 to 16, in increments of one thread. Similarly, we tested the MPI-based implementations by varying number of MPI processes, each bound to a different cluster node. In this way we tested MPI-based implementations on 2, 4, 8, 12, 16, 20, 24, 28 and 32 computing nodes. Note that varying the number of processing elements is not applicable to the single-GPU implementations of the programs, as they always use all available processing resources on a GPU. Only one of the two 1D and 2D programs were tested (corresponding to x direction in 1D and x-y plane in 2D), because, performance-wise, there is no difference between them (e.g., imag1dX-th vs. imag1dZ-th).

The main performance indicator is the execution time, the wall-clock time of one iteration of the main loop, averaged over 5 executions of 1000 iterations, reported in milliseconds. Using the results obtained, we calculated the speedup of all programs compared to the published serial C implementation. We were also interested in examining the scaling efficiency, or *scalability*, of the OpenMP and MPI programs. We tested both *strong* scaling, when the mesh size stays the same but the number of processing elements varies, and *weak* scaling, when the amount of work each processing element performs stays the same while the number of processing elements increases. More formally, given the execution time of a single iteration of serial programs ($T(1)$), and the corresponding execution time for parallel programs performed with N processing elements ($T(N)$), we calculated speedup as $S(N) = T(1)/T(N)$ and strong scaling efficiency as $E(N) = S(N)/N$. Weak scaling is computed as $E_W(N) = T_W(1)/T_W(N)$, where $T_W(1)$ is the execution time of a program using single processing element performing the work assigned to it, while $T_W(N)$ is the execution time of a program using N processing elements performing N times more work. We achieve this by increasing the mesh size.

2.4.3. Performance Test Results and Modeling of Single Node Programs

In this section we present the results obtained for single computing node OpenMP, CUDA and hybrid programs, and compare them to the previously published [23] serial implementation.

Table 2. Performance testing matrix, showing the mesh sizes and numbers of processing elements (threads or processes) used to test the programs, as well as the baseline program used for comparison

Program	Mesh size		Processing elements		Baseline
	Min	Max	Min	Max	
OpenMP programs					
imag1dX-th	1000	1000000	1	16	imag1d
real1dX-th	1000	1000000			real1d
imag2dXY-th	1000 × 1000	15000 × 15000	1	16	imag2dXY
real2dXY-th	1000 × 1000	13000 × 13000			real2dXY
imag3d-th	50 × 50 × 50	800 × 800 × 800	1	16	imag3d
real3d-th	50 × 50 × 50	800 × 800 × 800			real3d
CUDA programs					
imag2dXY-cuda	1000 × 1000	15000 × 15000	1	1	imag2dXY
real2dXY-cuda	1000 × 1000	13000 × 13000			real2dXY
imag3d-cuda	50 × 50 × 50	600 × 600 × 600	1	1	imag3d
real3d-cuda	50 × 50 × 50	540 × 540 × 540			real3d
Hybrid programs					
imag2dXY-hetero	1000 × 1000	15000 × 15000	16+1	16+1	imag2dXY-th
real2dXY-hetero	1000 × 1000	13000 × 13000			real2dXY-th
imag3d-hetero	50 × 50 × 50	600 × 600 × 600	16+1	16+1	imag3d-th
real3d-hetero	50 × 50 × 50	600 × 600 × 600			real3d-th
OpenMP/MPI programs					
imag3d-mpi	480 × 480 × 250	1920 × 1920 × 960	1×16	32×16	imag3d-th
real3d-mpi	480 × 480 × 250	1920 × 1920 × 960			real3d-th
CUDA/MPI programs					
imag3d-mpicuda	480 × 480 × 250	1920 × 1920 × 960	1	32	imag3d-cuda
real3d-mpicuda	480 × 480 × 250	1920 × 1920 × 960			real3d-cuda
Hybrid/MPI programs					
imag3d-mpihetero	480 × 480 × 250	1920 × 1920 × 960	1×(16+1)	32×(16+1)	imag3d-hetero
real3d-mpihetero	480 × 480 × 250	1920 × 1920 × 960			real3d-hetero

Table 3. Wall-clock execution times of a single iteration of the main time-propagation loop of single-node OpenMP programs (in milliseconds) for different number of OpenMP threads $N_{\rm th}$ and speedup $S(16)$ in strong scaling tests. The speedup is calculated w.r.t. the execution times of previously published serial versions of programs [23], given in the second column

Program	Serial	$N_{\rm th}=1$	$N_{\rm th}=2$	$N_{\rm th}=4$	$N_{\rm th}=8$	$N_{\rm th}=16$	$S(16)$
imag1dX-th	9.1	7.1	4.7	3.4	2.9	2.8	2.5
real1dX-th	15.2	14.2	10.5	8.2	7.3	7.2	2.0
imag2dXY-th	13657	7314	4215	2159	1193	798	9.2
real2dXY-th	17281	11700	6417	3271	1730	1052	11.1
imag3d-th	16064	9353	5201	2734	1473	888	10.5
real3d-th	22611	17496	9434	4935	2602	1466	11.9

Strong scaling performance test results for the OpenMP-based implementation using methodology described in the previous section are given in Table 3 and Figure 9. They show the obtained execution times, speedups and strong scaling efficiencies for different number of OpenMP threads. Columns $N_{\rm th}=1$, $N_{\rm th}=4$, $N_{\rm th}=8$ and $N_{\rm th}=16$ in Table 3 correspond to the number of threads used, while the last column shows the obtained speedup $S(16)$ with 16 OpenMP threads compared to one OpenMP thread. Strong scaling efficiency $E(N_{\rm th}) = S(N_{\rm th})/N_{\rm th}$ in Figure 9 is calculated as a fraction of the obtained speedup compared to a theoretical maximum. The mesh size used in 1D is 10^5, in 2D $10^4 \times 10^4$, while in 3D the mesh size is $480 \times 480 \times 480$. Execution times and speedups of imag1dZ-th, real1dZ-th, imag2dXZ-th, and real2dXZ-th (not reported here) are similar to those of imag1dX-th, real1dX-th, imag2dXY-th, and real2dXY-th, respectively.

The change from C2C to R2C FFT routine has a big impact on the execution time of single-threaded ($N_{\rm th} = 1$) programs compared to the previous serial programs. As we can see from the table, these improvements alone yield a speedup of 1.3 to 1.9 in 2D and 3D programs, and somewhat smaller speedup for 1D programs, 1.1 to 1.3. The use of additional threads brings about further speedup (reported in the last column) of 2 to 2.5 for 1D programs, and 9 to 12 for 2D and 3D programs. In Figure 9 we see that the efficiency rapidly decreases for 1D programs, even though speedup increases with the number of threads used.

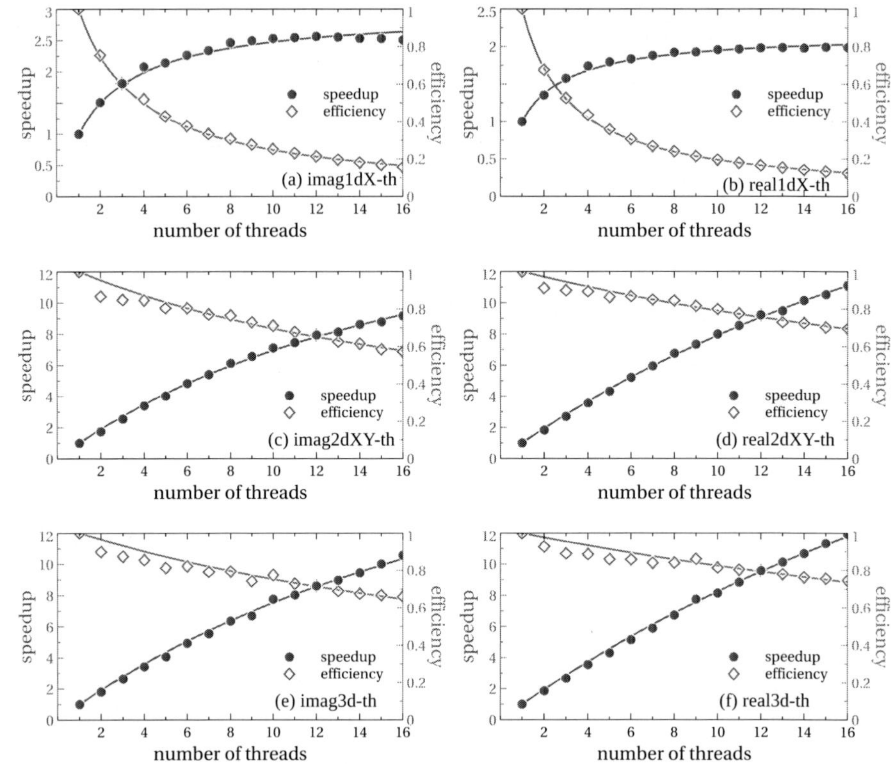

Figure 9. Speedup in the execution time and strong scaling efficiency of OpenMP programs compared to single-threaded runs for: (a) imag1dX-th, (b) real1dX-th, (c) imag2dXY-th, (d) real2dXY-th, (e) imag3d-th, (f) real3d-th. Solid lines represent fits to measured data, where fit model functions are given in the text and obtained fit parameters are listed in Table 4. Note that the model parameter p is fitted only to the speedup data, and then used to plot the efficiency model curve.

This is expected, as parts of the algorithm dealing with the recursive relations for calculation of the CN coefficients are inherently serial. In 1D, already with $N_{\rm th} = 4$ threads we almost achieve the maximal speedup, while still keeping the efficiency around 50%. We also see that, as expected, speedup and efficiency of multidimensional programs behave quite well as we increase the numbers of threads. In particular, we note that the efficiency always remains above 60%, making the use of all available CPU cores worthwhile.

From Figure 9 we observe that the speedup of 1D programs saturate quickly due to inherent serial nature of the portion of the algorithm, while in 2D and 3D the speedup behaves almost linearly. Despite their obvious differences, all curves in Figure 9 can be successfully modeled based on Amdahl's law [58]. Namely, the measured execution time $T(N_{\text{th}})$ of one iteration of the main loop can be expressed as

$$T(N_{\text{th}}) = T(1)\left(s + \frac{p}{N_{\text{th}}}\right), \tag{33}$$

where N_{th} is the number of threads used, $T(1)$ is the execution time of a single-threaded run, s is the serial fraction of the loop code, and p is the corresponding parallel fraction. By definition, $s + p = 1$, and therefore the speedup can be modeled by

$$S(N_{\text{th}}) = \frac{T(1)}{T(N_{\text{th}})} = \frac{1}{1 - p + p/N_{\text{th}}}, \tag{34}$$

where the parallel fraction of the main loop code p is the only fit parameter. Table 4 gives the obtained model fit parameter values for the data from Figure 9. As we can see, the fits match the obtained measurement data very well. Note that the efficiency can be expressed in terms of the speedup model

$$E(N_{\text{th}}) = \frac{S(N_{\text{th}})}{N_{\text{th}}} = \frac{1}{N_{\text{th}}}\frac{1}{1 - p + p/N_{\text{th}}}, \tag{35}$$

and is not fitted independently. Figure 9 shows that this model with the parameter p fitted on the speedup data matches very well with the efficiency data points.

From the obtained values of the parallel fraction of the algorithm, we see that 2D and 3D programs are almost ideally parallelizable, with p over 95%. In 1D case, however, the parallel fraction is around 66% for imaginary-time propagation and around 54% for real-time propagation, due to the fact that calculation of CN coefficients in the function `calclux` cannot be parallelized. Note that the parallel fraction of imaginary-time 1D program is higher than that of the real-time 1D program due to the larger amount of arithmetic operations required to process complex-valued data in the real-time version of the serial function `calclux`.

OpenMP programs were also tested for weak scalability. We were mostly interested in 3D variants of the programs, which we tested by fixing the amount

Table 4. Values of obtained strong scaling model fit parameter p and the estimated fit errors for OpenMP-parallelized programs

Program	p	Δp
imag1dX-th	0.662	0.004
real1dX-th	0.541	0.003
imag2dXY-th	0.9514	0.0007
real2dXY-th	0.9706	0.0004
imag3d-th	0.9635	0.0008
real3d-th	0.9762	0.0006

of work to 6,912,000 spatial points of the mesh, which corresponds to a mesh size of $240 \times 240 \times 120$. By increasing the number of OpenMP threads, we also increase the mesh size to be the multiple of 6,912,000. This means that the mesh has to be increased in such a way that, when divided among the threads, each thread gets 6,912,000 spatial points to process. The mesh sizes that satisfy this requirement cannot always be obtained by multiplying all three array dimensions with the same number, so we have to work with mesh sizes in which the size of each dimension may be different. For such mesh sizes, we tested all possible combinations, however no significant difference has been observed. Table 5 contains mesh sizes we used for testing for $N_{\rm th} = 1, 2, 4, 8, 12, 16$ threads, along with the execution times (in milliseconds) and the weak scaling efficiency. The testing was also done for $N_{\rm th} = 6$ and 10 threads, but the corresponding mesh sizes are omitted from the table for brevity. However, Figure 10 presents complete data collected in this test, for all $N_{\rm th}$ values, averaged over mesh sizes used. The figure shows that the real-time 3D programs have better weak scaling efficiency, which is about 75% at 16 threads, while imaginary-time programs demonstrate smaller, but still significant efficiency of about 60%.

To model weak scaling efficiency, we compare execution times of a single iteration of the main loop $T_W(N_{\rm th})$, performed with $N_{\rm th}$ threads, where for each value of $N_{\rm th}$ the total amount of work is $N_{\rm th}$ times the work being performed in a single-threaded run, i.e., the amount of work per thread is constant. The weak scaling efficiency is defined as

$$E_W(N_{\rm th}) = \frac{T_W(1)}{T_W(N_{\rm th})}, \tag{36}$$

where $T_W(1) = T(1)$ is the execution time of a single-threaded run for a given

Table 5. Weak scaling efficiency of OpenMP-parallelized 3D programs. Wall-clock execution times T_W are given in milliseconds, efficiencies E_W in percents

Mesh size	imag3d-th		real3d-th	
	T_W	E_W	T_W	E_W
$N_{\mathrm{th}} = 1$				
$120 \times 240 \times 240$	699	98.2	1259	99.5
$240 \times 120 \times 240$	711	96.5	1271	100
$240 \times 240 \times 120$	686	100	1275	98.2
$N_{\mathrm{th}} = 2$				
$240 \times 240 \times 240$	718	95.5	1254	99.9
$N_{\mathrm{th}} = 4$				
$240 \times 240 \times 480$	755	90.9	1312	95.5
$240 \times 480 \times 240$	778	88.2	1301	96.3
$480 \times 240 \times 240$	754	91	1310	95.6
$N_{\mathrm{th}} = 8$				
$240 \times 480 \times 480$	875	78.4	1405	89.9
$480 \times 240 \times 480$	847	81	1393	89.1
$480 \times 480 \times 240$	886	77.5	1431	87.7
$N_{\mathrm{th}} = 12$				
$360 \times 480 \times 480$	1010	67.9	1532	81.7
$480 \times 360 \times 480$	981	69.9	1514	82.7
$480 \times 480 \times 360$	1004	68.4	1581	79.2
$N_{\mathrm{th}} = 16$				
$480 \times 480 \times 480$	1155	59.4	1680	74.5

workload (in our case, 6,912,000 spatial points). The expected execution time of a workload assigned to a simulation with N_{th} threads, executed in a single-threaded run, is $N_{\mathrm{th}}T(1)$. In weak scaling tests, this workload is executed with N_{th} threads and therefore the expected execution time can be modeled by

$$T_W(N_{\mathrm{th}}) = N_{\mathrm{th}}T(1)\left(1 - p + \frac{p}{N_{\mathrm{th}}}\right). \tag{37}$$

According to this argument, we model the weak scaling efficiency by a single-parameter function

$$E_W(N_{\mathrm{th}}) = \frac{1}{p + (1-p)N_{\mathrm{th}}}. \tag{38}$$

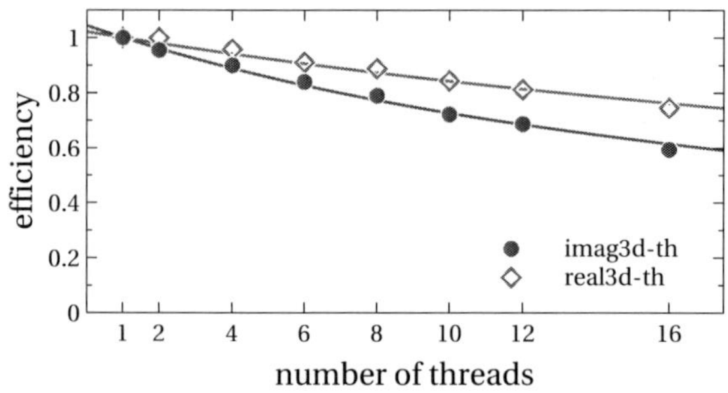

Figure 10. Weak scaling efficiency of OpenMP-parallelized 3D programs, averaged over all mesh sizes tested for each value of $N_{\rm th}$. Solid lines represent fits to measured data, where fit model functions are given in the text.

We fitted this model to data presented in Figure 10 and the obtained fit parameters are given in Table 6. As we can see from the figure, the above model is an excellent fit to experimental data, and both 3D OpenMP-parallelized programs have high (above 95%) parallel fraction of the code, in agreement with the results of strong scaling tests (Table 4).

Table 6. Values of obtained weak scaling model fit parameter p and the estimated fit errors for OpenMP-parallelized programs

Program	p	Δp
imag3d-th	0.958	0.002
real3d-th	0.9792	0.0009

Overall, we conclude that imaginary-time OpenMP-parallelized programs show smaller speedup and efficiency across all tests, except in 1D. This can be attributed to the additional step of wave function normalization in each iteration of the imaginary-time propagation, as well as the fact that real-time programs work with complex-valued data that require more arithmetic operations for the same mesh size. Since much of the computation inside loops requires simple arithmetic, the throughput of the CPU is often not fully exploited in imaginary-time programs, thus the pressure of memory bandwidth makes these programs less efficient. The real-time programs are also affected by this, however to a somewhat lesser extent.

Next, we consider the CUDA implementation of the shared memory algorithm. GPU functions as a single processing element, therefore we cannot test CUDA implementation by varying both the mesh size and the number of processing elements. However, just varying the mesh size gives us valuable insight into the behavior of this implementation, due to the difference in programming models and the libraries used. Tables 7 and 8 show the execution times (in milliseconds) for a number of mesh sizes tested, as well as the average speedup $\langle S \rangle$ compared to the serial programs [23]. Figure 11 shows the speedup obtained for all mesh sizes tested, and horizontal lines represent average speedups obtained for each program. Note that the dispersion of data is due to the use of `FFTW_ESTIMATE` flag in library calls to FFTW in the serial programs. Use of this flag results in a choice of suboptimal FFT algorithm for some mesh sizes. The vertical lines in Figure 11 denote the change in `POTMEM` parameter. The speedups left of the first vertical line are obtained with `POTMEM=2` and thus demonstrate the best speedup. Second group of results, between the two vertical lines, is obtained with `POTMEM=1`, and we note that the speedup decreases slightly, while the results right of the second vertical line are obtained with `POTMEM=2`, and show the smallest speedup due to the use of mapped memory. The 2D programs in x-z plane exhibited very similar performance to those of x-y plane, and therefore we did not include them in the figures.

Table 7. Wall-clock execution times of a single iteration of the main time-propagation loop of single-node 2D CUDA programs (in milliseconds) for different mesh sizes, and average speedup w.r.t. to the execution times of serial 2D programs [23]

Program	2000^2	4000^2	6000^2	8000^2	10000^2	12000^2	Avg. speedup
imag2dXY-cuda	24.1	104.3	235.2	386.1	657.1	1150.4	10
real2dXY-cuda	29.9	112.4	266.4	444.0	749.0	1528.3	14

For small mesh sizes, the GPU remains underutilized, resulting in a smaller speedup. For large mesh sizes, where the GPU memory usage approaches the limit, we also see declining speedup. This is due to the inevitable use of `POTMEM` parameter, which keeps some arrays in the host memory when they cannot fit in the memory of the GPU. Overall, the CUDA implementation shows execution times similar to the OpenMP implementation for imaginary-time propagation, and slightly lower for real-time propagation. We stress that these results strongly depend on the type of GPU used and that the obtained speedups

Table 8. Wall-clock execution times of a single iteration of the main time-propagation loop of single-node 3D CUDA programs (in milliseconds) for different mesh sizes, and average speedup w.r.t. to the execution times of serial 3D programs [23]

Program	100^3	200^3	300^3	400^3	500^3	Avg. speedup
imag3d-cuda	10.6	79.3	298.8	674.5	1260.2	7.1
real3d-cuda	10.9	84.1	302.5	682.4	1467.4	13.5

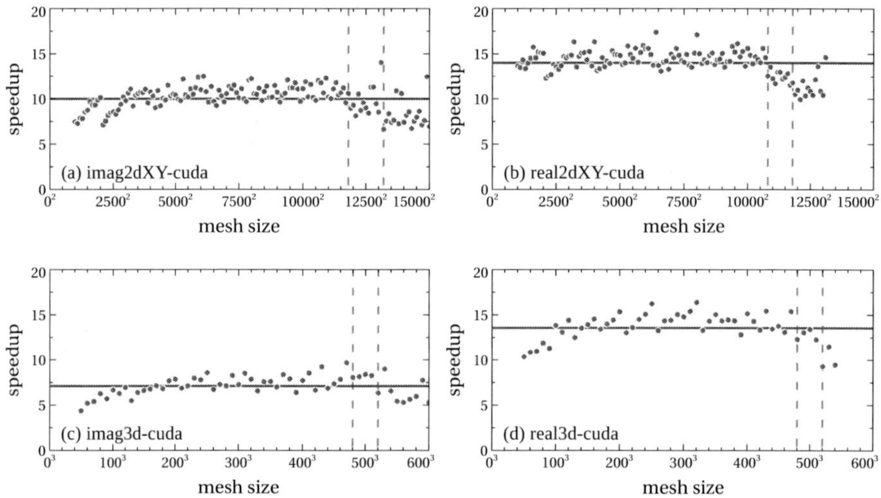

Figure 11. Speedup in the execution time of CUDA programs compared to the serial programs for all tested mesh sizes: (a) imag2dXY-cuda, (b) real2dXY-cuda, (c) imag3d-cuda, (d) real3d-cuda. Horizontal lines represent $\langle S \rangle$, while dashed vertical lines represent different values of `POTMEM` parameter (see text).

may be even better for newer-generation GPUs (e.g., *Kepler-*, *Maxwell-*, *Pascal-* and *Volta*-based GPUs).

Hybrid OpenMP/CUDA implementation was tested on a range of mesh sizes, similarly to the OpenMP- and CUDA-based implementations. Since OpenMP implementation showed that all cores should be used for 2D and 3D programs, we kept the number of OpenMP threads fixed at 16 in our hybrid algorithm. The parameters governing the amount of data offloaded to GPU were optimized using our GA method from section 2.4.1 An illustrative subset of the results for 3D programs is shown in Table 9, which gives execution times for a single iteration of the main loop, for different mesh sizes. A comparison of execution times of hybrid and pure OpenMP programs with $N_{\rm th} = 16$ threads is given in Figure 12(a), while Figure 12(b) compares the performance of hybrid and pure CUDA programs, for all tested mesh sizes.

Table 9. Wall-clock execution times of a single iteration of the main time-propagation loop of single-node 3D hybrid programs (in milliseconds) for different mesh sizes. Offload parameters are optimized using the GA optimization method

	80^3	200^3	320^3	440^3	560^3	680^3	800^3
imag3d-hetero	6.3	67.1	271.4	678.9	1493.3	2750.9	4461.6
real3d-hetero	8.0	94.7	374.1	922.8	1985.2	4020.2	6676.6

Figure 12. Speedup in the execution time of 3D hybrid programs compared to: (a) OpenMP programs with $N_{\rm th} = 16$ threads, (b) CUDA programs. Dashed horizontal lines correspond to a speedup value $S = 1$.

From Figure 12 we see that the hybrid implementation outperforms the OpenMP one for all mesh sizes except for the smallest one. The same applies to the comparison of imaginary-time hybrid and CUDA programs, while real-time hybrid program outperforms the corresponding CUDA program only for mesh sizes larger than $400 \times 400 \times 400$. Although one would expect that the speedup of the optimized hybrid algorithm is always equal to or larger than one, we see that this is not the case for all mesh sizes in Figure 12. This is due to the fact that the hybrid FFT algorithm is always employed in hybrid programs, meaning that even if no or all data are offloaded to GPU, the splitting of FFT will still take place. Therefore, in these limiting cases we do not obtain pure OpenMP or CUDA algorithms that would yield maximal performance. Splitting the computation of FFT along one direction in `calcpsidd2` function disables some of the potential optimizations that libraries like FFTW and cuFFT exploit, and introduces data copies between host and device which cannot be fully offset using CUDA streams. The hybrid algorithm can compensate for this if the amount of offloaded data to GPU is sufficiently large, which is not the case for the smallest mesh size.

Table 10. Optimal fraction of total data offloaded to GPU for different mesh sizes in hybrid 3D programs, obtained using the GA optimization method

Mesh size	GPU portion of data [%]							
	imag3d-hetero				real3d-hetero			
	PS_1	PS_2	PS_3	PS_4	PS_1	PS_2	PS_3	PS_4
$80 \times 80 \times 80$	20	3	60	33	23	3	40	10
$120 \times 120 \times 120$	12	10	25	28	23	10	25	27
$160 \times 160 \times 160$	15	7	31	24	44	19	34	23
$200 \times 200 \times 200$	38	27	37	28	44	26	46	26
$240 \times 240 \times 240$	38	28	40	30	38	28	47	25
$280 \times 280 \times 280$	40	31	30	33	46	22	37	25
$320 \times 320 \times 320$	33	38	29	36	45	28	69	44
$360 \times 360 \times 360$	41	33	27	36	49	33	43	31
$400 \times 400 \times 400$	38	25	30	38	46	24	47	21
$440 \times 440 \times 440$	31	35	29	39	45	35	43	32
$480 \times 480 \times 480$	16	19	22	41	45	33	62	35
$520 \times 520 \times 520$	40	31	28	46	45	25	40	37
$560 \times 560 \times 560$	38	41	30	43	36	24	35	32
$600 \times 600 \times 600$	37	27	30	43	29	29	29	28
$640 \times 640 \times 640$	36	40	30	43	24	22	24	23
$680 \times 680 \times 680$	26	38	28	38	20	16	20	20
$720 \times 720 \times 720$	23	26	25	32	17	13	17	17
$760 \times 760 \times 760$	21	22	25	27	14	14	14	13
$800 \times 800 \times 800$	19	23	23	21	11	8	12	12

Given that pure CUDA programs performing imaginary-time propagation demonstrated smaller speedup than the corresponding OpenMP programs (Figure 11 vs. Figure 9), it is expected that CUDA portion of the hybrid imaginary-time propagation programs would yield smaller improvement, which is evident when we compare hybrid implementation with the OpenMP one in Figure 12(a), except for the largest mesh sizes. We can reach the same conclusion if we consider that, in real-time propagation, CUDA implementation shows the highest speedup of all single-node programs, and therefore the corresponding hybrid programs show better speedup than imaginary-time propagation programs parallelized with OpenMP. On the other hand, when compared with the CUDA implementation in Figure 12(b), the situation is reversed, and we see much better performance of the hybrid imaginary-time propagation programs, while the hybrid real-time propagation programs achieve speedup larger than one only for mesh sizes larger than $400 \times 400 \times 400$. This result can be attributed to the fact that in real-time propagation the amount of data copied between host and device is larger, given that some of the copied arrays are complex-valued.

Also, we can observe that the speedup in Figure 12(a) declines with mesh size. This is due to the memory saturation of the GPU device, as only a small portion of the data can be offloaded when the mesh is large. The amount of data offloaded to GPU for all tested mesh sizes is given in Table 10, where we can see that the total amount of data processed by the GPU declines with increasing mesh size. As a result, the computation is unbalanced, given that the host has to work with a much larger portion of the data. This is reflected in the optimization step, where our GA quickly converges to the highest possible mesh size that can be offloaded to the device, implying that a greater amount of device memory would be required to get a balanced computation.

To make an optimal choice of programs to be used on a single node for a particular hardware platform, one has to perform detailed tests using the methodology presented in section 2.4.2 The results presented in this section apply to hardware comparable to the one available at PARADOX supercomputing facility, but one can expect similar behavior for all modern types of CPU and GPU.

2.4.4. Performance Test Results and Modeling of MPI Programs

Following on from the previous section, here we present the results obtained for MPI-parallelized programs, executed on a computer cluster consisting of varying number of computing nodes, and compare them with the corresponding single-node programs.

MPI programs are highly dependent on the configuration of the cluster, mainly on the speed of interconnect, but also on the distribution of processes and threads, NUMA configuration, MPI configuration, etc. Getting the best performance out of the programs requires some experimentation with several different configurations. The results presented are obtained without extensive tuning of the cluster or MPI runtime, with the aim to show the base performance.

Strong scaling of MPI-based implementations is tested in a similar way to the single-node testing, by varying number of cluster nodes N_{p} from 2 to 32. On each node, a single MPI process is launched, which then uses the node's resources, either by further spawning 16 OpenMP threads in case of OpenMP/MPI and Hybrid/MPI implementations, or by invoking the CUDA kernels utilizing the node's GPU in case of CUDA/MPI implementation. In this test, the baseline used for comparison was the equivalent single-node implementation, i.e., OpenMP/MPI programs are compared with OpenMP programs executing with $N_{\mathrm{th}} = 16$ threads, CUDA/MPI programs are compared to the single-node pure CUDA programs, and Hybrid/MPI programs is compared to the single-node hybrid programs.

Mesh size we use in this test is $480 \times 480 \times 250$. Since the values of `Nx` and `Ny` parameters must be divisible by the number of processes N_{p}, this mesh cannot be distributed over 28 processes. In this case we use slightly modified mesh, taking into account the multiple of 28 that is closest to 480, yielding a mesh size of $476 \times 476 \times 250$. Data is never distributed along z direction, so no such requirement exists for `Nz`. The resulting mesh has slightly fewer spatial points, and thus less work per process, potentially allowing for a better performance, but in our tests the smaller mesh did not have a significant impact on the measured execution time.

Note that strong scaling tests, where the mesh size is fixed, inevitably lead to saturation and decrease in measured speedup and efficiency values. Whatever the chosen mesh size, increase in the number of cluster nodes used will eventually yield insufficient amount of work per MPI process, such that communication will start to dominate over computation. Therefore, in our strong scaling tests we can expect that execution times initially decrease with increasing values of N_{p}, but eventually performance of MPI-based programs will decline. This issue is addressed by considering weak scaling tests.

Table 11. Wall-clock execution times of a single iteration of the main time-propagation loop of OpenMP/MPI, CUDA/MPI and Hybrid/MPI programs (in milliseconds) for different number of MPI processes N_p and speedup $S(32)$ in strong scaling tests. The speedup is calculated w.r.t. the baseline execution times of the corresponding single-node programs (OpenMP, CUDA and hybrid, respectively), given in the second column.

Program	Baseline	$N_\mathrm{p} = 4$	$N_\mathrm{p} = 8$	$N_\mathrm{p} = 16$	$N_\mathrm{p} = 24$	$N_\mathrm{p} = 32$	$S(32)$
imag3d-mpi	1124	541	262	134	89	64	17.5
real3d-mpi	2140	700	358	207	155	98	21.8
imag3d-mpicuda	579	438	210	103	71	59	9.8
real3d-mpicuda	800	609	291	142	95	79	10.1
imag3d-mpihetero	489	299	162	99	84	81	6.0
real3d-mpihetero	613	407	255	154	135	101	6.0

The illustrative portion of obtained execution times of a single iteration of the main loop for the three MPI-parallelized implementations are shown in Table 11, together with the execution times of the corresponding baseline single-node program. Columns $N_\mathrm{p} = 4$, $N_\mathrm{p} = 8$, $N_\mathrm{p} = 16$ and $N_\mathrm{p} = 32$ correspond to the number of cluster nodes used, while the last column shows the obtained speedup $S(32)$ with $N_\mathrm{p} = 32$ nodes compared to baseline single-node programs.

The maximal speedup of OpenMP/MPI implementation ranges from 17 for imaginary-time propagation to 22 for real-time propagation programs for $N_\mathrm{p} = 32$. The complete measurement results for speedup and efficiency are depicted in Figure 13(a) and 13(b), where we see that the speedup grows linearly with the number of nodes used, while the efficiency remains mostly constant in the range between 40% and 60%, thus making the use of OpenMP/MPI programs highly advantageous for simulations with large mesh size. In general, we can expect even better efficiency for larger mesh sizes.

Similar behavior is observed for CUDA/MPI implementation. The obtained speedup with $N_\mathrm{p} = 32$ nodes ranges from 9 to 10, with the slightly lower efficiency, between 30% and 40%, as shown in Figure 13(c) and 13(d). Even though the efficiency is lower for this implementation, the speedup still grows linearly and the execution times are lower than for the OpenMP/MPI implementation. This makes CUDA/MPI programs ideal choice for use on GPU-enabled computer clusters. Additional benefit of using CUDA/MPI programs is their low CPU usage (using only one CPU core per cluster node), allowing for the possibility that the same cluster nodes are used for other CPU-intensive simulations in a time-sharing fashion.

As we see in Figure 13(e) and 13(f), the linear growth of speedup is also present for the Hybrid/MPI implementation, however, as the amount of work per-process shrinks, the efficiency drastically drops down to 20% with $N_\mathrm{p} = 32$ nodes. For the mesh size used in this test, we observe that the Hybrid/MPI implementation performs very well on $N_\mathrm{p} < 16$ processes, providing the lowest execution times, but with $N_\mathrm{p} \geq 16$ processes its execution times become larger than for the CUDA/MPI implementation (especially for real-time propagation programs), and eventually even for the OpenMP/MPI implementation, as illustrated in Figure 14. This is again due to the insufficient amount of work each computing node performs after data are distributed among the MPI processes, which can be seen from the amount of data offloaded to GPU for all tested values of N_p in Table 12. Similar saturation in performance will eventually happen

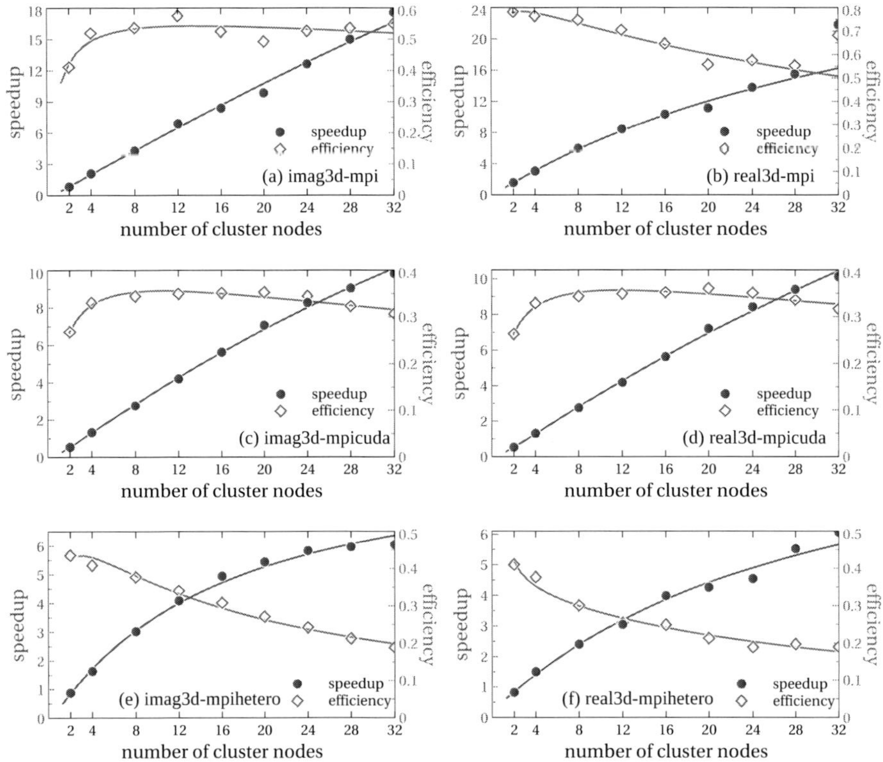

Figure 13. Speedup in the execution time and strong scaling efficiency of MPI-based programs compared to single-node runs: (a) imag3d-mpi, (b) real3d-mpi, (c) imag3d-mpicuda, (d) real3d-mpicuda, (e) imag3d-mpihetero, (f) real3d-mpihetero. Solid lines represent fits to measured data, where fit model functions are given in the text and obtained fit parameters are listed in Table 13. Note that the model parameters are fitted only to the speedup data, and then used to plot the efficiency model curve.

for the OpenMP/MPI and CUDA/MPI implementations for larger values of N_{p}.

We thus conclude that Hybrid/MPI implementation has the best performance of the three MPI-based implementations if the amount of work per process remains high enough to justify the use of hybrid algorithm. Otherwise, either of the other two MPI-based implementations should be considered first. The energy efficiency of this and other MPI-based implementations was not

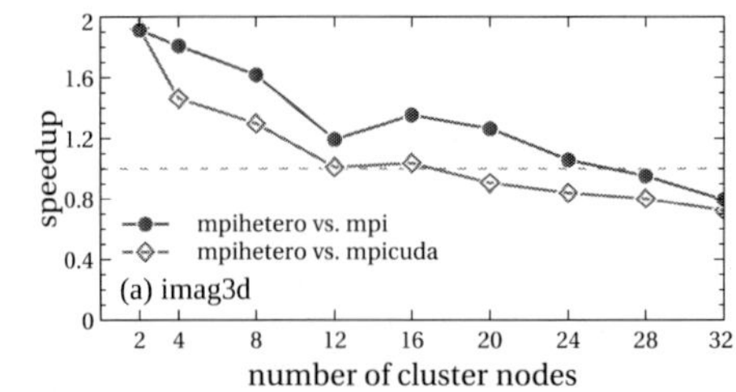

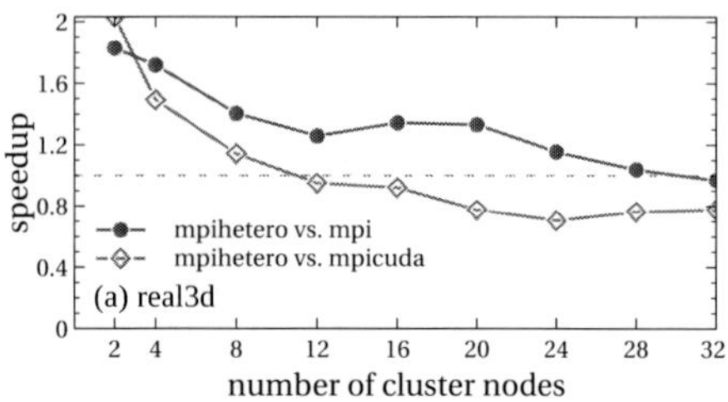

Figure 14. Speedup in the execution time of Hybrid/MPI programs in strong scaling tests compared to the other two MPI-based programs: (a) OpenMP/MPI, (b) CUDA/MPI.

explored due to the difficulty in making precise measurements. If the energy consumption is not an issue, Hybrid/MPI implementation will yield the best performance, providing the cluster has powerful GPUs installed and the mesh size used is large enough.

Table 12. Optimal fraction of total data offloaded to GPU per process for different values of N_p in Hybrid/MPI programs, obtained using the GA optimization method in strong scaling tests

N_p	GPU portion of data [%]							
	imag3d-mpihetero				real3d-mpihetero			
	PS_1	PS_2	PS_3	PS_4	PS_1	PS_2	PS_3	PS_4
1	30	17.5	34.4	36.7	45.2	32.5	55	35
2	25	10	17.5	16.7	20.8	11.5	23.8	16
4	10	5.2	9.2	9.4	10	5.2	15	6.7
8	5	2.9	5.8	6.7	6.3	2.1	6.7	5
12	3.3	1.7	4.2	4.2	4	1.3	5	3.3
16	2.9	1.3	2.9	2.1	3.1	0.4	2.9	2.1
20	1.7	0.6	2.5	1	2.5	0.4	2.5	1.7
24	1.7	1.3	2.1	1.7	2.1	1.3	2.5	0.8
28	1.7	0.4	2.1	1.3	1.7	0.4	1.7	1.3
32	0.8	0.4	1.7	0.8	1.5	0.4	2.1	0.8

We now model the speedup and strong scaling efficiency of MPI-based programs. In general, the execution time of a program can be expressed as

$$T(N_\mathrm{p}) = \alpha + \beta L + \gamma \frac{M}{N_\mathrm{p}} + \beta V \frac{M}{N_\mathrm{p}^2}\,, \tag{39}$$

where α represents the average time to perform serial portion of the code, L is the communication latency associated with one MPI message, β is the frequency of MPI messaging, γ is the average time to process one spatial point, V is related to data transfer speed (throughput), and M is the mesh size ($\mathtt{Nx} \times \mathtt{Ny} \times \mathtt{Nz}$). The communication overhead of one all-to-all message passing instance is equal to $L + VM/N_\mathrm{p}^2$, where each of N_p processes communicates its M/N_p part of the mesh evenly to all other processes, leading to a message size of M/N_p^2. Taking into account that in the strong scaling tests the mesh size is fixed, the above model can be simplified to

$$T(N_\mathrm{p}) = T(1)\left(a + \frac{b}{N_\mathrm{p}} + \frac{c}{N_\mathrm{p}^2}\right)\,, \tag{40}$$

while the speedup can be modeled by

$$S(N_\mathrm{p}) = \frac{T(1)}{T(N_\mathrm{p})} = \frac{1}{a + b/N_\mathrm{p} + c/N_\mathrm{p}^2}\,. \tag{41}$$

Table 13. Values of obtained strong scaling model fit parameters a, b and c, as well as the estimated fit errors for MPI-parallelized programs

Program	a	Δa	b	Δb	c	Δc
imag3d-mpi	0.008	0.006	1.6	0.2	1.4	0.4
real3d-mpi	0.026	0.006	1.1	0.1	0.2	0.3
imag3d-mpicuda	0.025	0.008	2.3	0.2	2.8	0.6
real3d-mpicuda	0.020	0.008	2.3	0.2	2.8	0.6
imag3d-mpihetero	0.104	0.009	1.7	0.2	0.9	0.5
real3d-mpihetero	0.09	0.02	2.7	0.3	-1.0	0.6

This model is fitted to the obtained strong scaling measurement data and the results are presented in Table 13, while the corresponding model curves are shown as solid lines in Figure 13. As we see, the proposed model agrees very well with the experimental data. The only exception is the speedup and efficiency of real-time OpenMP/MPI program (real3d-mpi) on $N_\mathrm{p} = 32$ processes, where FFTW library creates an optimal transform plan that works very well with the given mesh size. However, changing the mesh size even slightly gives the performance comparable to the model prediction. We note that the value of the model parameter c is negative for real-time Hybrid/MPI program (real3d-mpihetero), which should not be the case since is parameter is related to data transfer speed. However, due to uncertainty of the optimization choice by the GA method and the fact that the fraction of total data offloaded to GPU gradually decreases (Table 12), leading from real hybrid algorithm to almost pure OpenMP/MPI one with hybrid FFT, it is not surprising that the obtained value differs from expected. Taking into account relatively large fit error $\Delta c/c = 60\%$ for this parameter, we can still use expression (41) to model performance of Hybrid/MPI programs.

Next, we test weak scaling of MPI-based implementations. The same number of cluster nodes (and thus MPI processes) was used as in previous tests, while the starting mesh, which corresponds to a unit workload, had a size of $480 \times 480 \times 480$, amounting to 110,595,000 spatial points. This number of spatial points was kept constant per process, similarly to the weak test of OpenMP programs. There is an exception to this scheme, the case when programs are executed on $N_\mathrm{p} = 28$ nodes, as the scaled number of spatial points cannot be evenly distributed among 28 processes. In this case we compare the weak scalability by scaling up a starting mesh of $476 \times 476 \times 476$, which has 107,850,176 spatial points, just slightly less than for other values of N_p. The base of comparison was the execution time of a single iteration of the main loop with the corresponding MPI-based program running as a single process ($N_\mathrm{p} = 1$). While we used this configuration as the baseline, we do not recommend running MPI-based programs with $N_\mathrm{p} = 1$ processes, because no special handling of such case has been implemented. This means that MPI-parallelized programs are always transposing the data, which is unnecessary with $N_\mathrm{p} = 1$ as all data is local to the single process. Instead, we recommend using single-node variants of the programs outside cluster environment.

Table 14. Weak scaling of MPI-based programs. Wall-clock execution times T_W are given in milliseconds, efficiencies E_W in percents

Mesh size	imag3d-mpi		real3d-mpi		imag3d-mpicuda		real3d-mpicuda		imag3d-mpihetero		real3d-mpihetero	
	T_W	E_W	T_W	E_W	T_W	E_W	T_W	E_W	T_W	E_W	T_W	E_W
$N_p = 1$												
$480 \times 480 \times 480$	1758	100	2956	100	4338	100	6346	100	1971	100	2877	100
$N_p = 2$												
$480 \times 480 \times 960$	3858	45.6	4910	60.2	3785	114.6	5430	116.9	2044	96.4	2878	100
$480 \times 960 \times 480$	3704	47.5	5035	58.7	3825	113.4	5573	113.9	1997	98.7	2928	98.3
$960 \times 480 \times 480$	3758	46.8	5506	53.7	3871	112.1	5609	113.1	2019	97.6	2980	96.5
$N_p = 4$												
$480 \times 960 \times 960$	4267	41.2	6033	49.2	2995	144.8	4278	148.3	2149	91.7	3009	95.6
$960 \times 480 \times 960$	3877	45.4	5223	56.6	3042	142.6	4318	146.9	2218	88.9	3274	87.9
$960 \times 960 \times 480$	4000	44	5063	58.4	3078	140.9	4424	143.4	2148	91.8	3106	92.6
$N_p = 8$												
$960 \times 960 \times 960$	4327	40.6	6285	47	2952	146.9	4167	152.3	2369	83.2	3338	86.2
$N_p = 12$												
$1440 \times 960 \times 960$	4400	40	6729	43.9	2902	149.5	4254	149.2	2335	84.4	3504	82.1
$960 \times 1440 \times 960$	4268	41.2	6263	47.2	2953	146.9	4167	152.3	2412	81.7	3440	83.6
$960 \times 960 \times 1440$	4234	40.2	6299	46.9	2865	151.4	4153	152.8	2413	81.7	3351	85.9
$N_p = 16$												
$1920 \times 960 \times 960$	4356	40.4	6419	46	3001	144.5	4278	148.3	2392	82.4	3434	83.8
$960 \times 1920 \times 960$	4266	41.2	6376	46.4	2970	146.1	4170	152.2	2552	77.2	3525	81.6
$960 \times 960 \times 1920$	4234	40.2	6508	45.4	2888	150.2	4148	153	2639	74.7	3606	79.8
$N_p = 24$												
$1440 \times 1920 \times 960$	4677	40	6895	42.9	2924	148.4	4266	148.8	2552	77.3	3770	76.3
$1440 \times 960 \times 1920$	4796	38.3	7076	41.8	2984	145.4	4235	149.8	2611	75.5	3699	77.8
$1920 \times 1440 \times 960$	4470	40.6	6439	45.9	3007	144.2	4286	148.1	2546	77.4	3504	82.1
$1920 \times 960 \times 1440$	4450	40.8	6425	46	2984	145.3	4251	149.3	2551	77.3	3550	81.1
$960 \times 1440 \times 1920$	4364	41.1	6846	43.2	2879	150.7	4105	154.6	2682	73.5	3725	77.2
$960 \times 1920 \times 1440$	4346	41	6806	43.4	2878	150.7	4133	153.5	2501	78.8	3562	80.8
$N_p = 32$												
$1920 \times 1920 \times 960$	4452	39.5	6704	44.1	3183	136.3	4518	140.5	2585	76.3	3715	77.5
$1920 \times 960 \times 1920$	4520	38.9	6802	43.4	3163	137.2	4480	141.7	2632	74.9	3769	76.3
$960 \times 1920 \times 1920$	4301	40.9	6663	44.4	3014	143.9	4351	145.9	2776	71	3933	73.0

Table 14 lists mesh sizes used and execution times obtained for $N_{\mathrm{p}} = 1$, 2, 4, 8, 12, 16 and 32 cluster nodes. We also tested weak scaling on $N_{\mathrm{p}} = 20, 24$ and 28, but have excluded them from the table for brevity. A complete comparison is shown in Figure 15. The ordering of mesh dimensions had a slightly greater impact than with single-node OpenMP programs, varying up to 10% for the OpenMP/MPI programs, and less for the other two MPI-based implementations. This means that for meshes of the same size, thus implying equal work, execution times were mostly lower for those with larger values of `Nx` (e.g., the execution time for a $960 \times 480 \times 960$ mesh is lower than for a $480 \times 960 \times 960$ one). However, no distinct pattern emerges that would give us a clue as to which order is the most favorable. Upon inspection, we find that the difference in execution times is due to the different FFT plans employed by the FFTW and cuFFT libraries. The communication time remains mostly the same, as is expected since the amount of data exchanged is the same.

The results of weak scaling tests warrant further discussion as the execution times and plotted efficiencies appear misleading. On one hand, OpenMP/MPI programs show poor scaling results, achieving only 40% efficiency, while on the other hand the CUDA/MPI show efficiency well above 100%, a seemingly impossible result. The lower efficiency of the OpenMP/MPI programs is due to the very good performance of the baseline run. Here, the FFTW creates a very good FFT and transpose plans, which exploit the fact that all data are local, and thus have very good execution time. With $N_{\mathrm{p}} = 2$ nodes the execution time increases significantly, and efficiency drops to about 45-55%. However, adding more work and nodes has much smaller impact on efficiency, which only drops to 40-45% with $N_{\mathrm{p}} = 32$ nodes. If we use $N_{\mathrm{p}} = 2$ as the baseline, we see that the efficiency remains above 80%, a very good result.

In contrast to the OpenMP/MPI version, CUDA/MPI version performs very badly when executed on one node. Due to message sizes, MPI implementation used (Open MPI) relies on asynchronous copies through host memory [59] to perform the required communication. This results in bad performance due to a lack of overlap when copying data, as there is only one process involved. With $N_{\mathrm{p}} = 2$ processes the situation improves, as multiple streams are used, much like how we employ streams to overlap computation and data transfers in implementation of the hybrid algorithm. Between $N_{\mathrm{p}} = 4$ and 16 processes we get the best performance, and further increasing the number of processes only slightly reduces efficiency due to the size and amount of messages passed. If we compare the weak scaling efficiency with the best result as the baseline

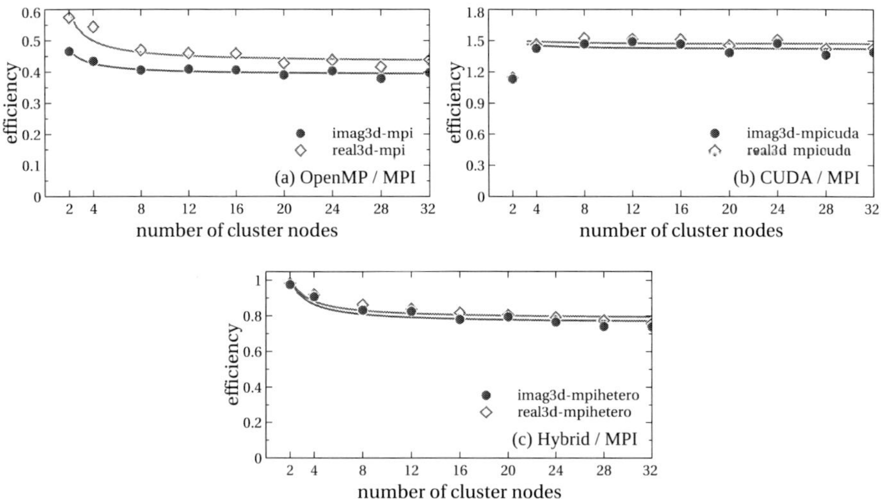

Figure 15. Weak scaling efficiency of three MPI-based implementations, averaged over all mesh sizes tested for each value of N_p. Solid lines represent fits to measured data, where fit model functions are given in the text.

(obtained with $N_\mathrm{p} = 16$ processes), we see that the efficiency is lower than for the OpenMP/MPI version, due to the different transpose routine used and divided multidimensional FFT that has to be employed. However, we stress that, in terms of absolute execution times, CUDA/MPI is faster than OpenMP/MPI version for $N_\mathrm{p} \geq 4$.

The Hybrid/MPI version behaves as expected and demonstrates weak scaling efficiency of about 70-75% with $N_\mathrm{p} = 32$ nodes. Since data are transposed only in host memory, this version does not suffer the penalty of memory copies like the CUDA/MPI version, and therefore achieves much better efficiency with small number of cluster nodes. In terms of absolute execution times, Hybrid/MPI version is the fastest of all three implementations for $N_\mathrm{p} \geq 2$ and is therefore the algorithm of choice for distributed memory systems. Of course, it assumes previous optimization of offloading parameters using the GA method, which itself is time-consuming and has to be taken into consideration when making the choice. Table 15 provides data on the optimal fraction of total data offloaded to GPU in our weak scaling tests. As the number of processes increases, one expects that the amount of data offloaded to GPU remains constant, which is the case, as we see from the table. The only exceptions are when the

Table 15. Optimal fraction of total data offloaded to GPU in weak scaling tests for different mesh sizes in Hybrid/MPI programs, obtained using the GA optimization method

Mesh size	imag3d-mpihetero				real3d-mpihetero			
	PS_1	PS_2	PS_3	PS_4	PS_1	PS_2	PS_3	PS_4
$N_p = 1$								
$480 \times 480 \times 480$	35	20	31	44	46	30	40	35
$N_p = 2$								
$480 \times 480 \times 960$	46	33	33	47	38	29	37	38
$480 \times 960 \times 480$	50	67	17	47	45	60	19	40
$960 \times 480 \times 480$	50	15	70	34	43	15	67	35
$N_p = 4$								
$480 \times 960 \times 960$	40	65	18	47	40	63	17	37
$960 \times 480 \times 960$	50	10	67	42	40	12	75	33
$960 \times 960 \times 480$	33	29	33	40	45	22	33	40
$N_p = 8$								
$960 \times 960 \times 960$	53	25	33	47	40	27	40	28
$N_p = 12$								
$1440 \times 960 \times 960$	37	20	53	45	42	18	60	63
$960 \times 1440 \times 960$	40	38	27	47	45	35	27	35
$960 \times 960 \times 1440$	50	25	38	45	45	25	40	40
$N_p = 16$								
$1920 \times 960 \times 960$	47	13	67	27	42	13	80	30
$960 \times 1920 \times 960$	47	47	35	37	47	53	20	25
$960 \times 960 \times 1920$	50	20	33	33	47	20	40	27
$N_p = 24$								
$1440 \times 1920 \times 960$	47	40	30	53	47	33	25	63
$1440 \times 960 \times 1920$	50	17	60	45	47	13	60	70
$1920 \times 1440 \times 960$	40	20	50	40	45	20	53	27
$1920 \times 960 \times 1440$	53	13	60	45	45	10	80	25
$960 \times 1440 \times 1920$	50	30	27	47	45	30	27	27
$960 \times 1920 \times 1440$	40	50	18	45	45	50	20	30
$N_p = 32$								
$1920 \times 1920 \times 960$	47	27	33	30	47	17	40	27
$1920 \times 960 \times 1920$	50	13	60	40	47	10	80	40
$960 \times 1920 \times 1920$	47	47	40	33	47	40	23	27

value of `Ny` is greater than `Nx` where we see a larger fraction of data offloaded to GPU in PS_2 and PS_4, and when the value of `Ny` is smaller than `Nx` where we see a smaller fraction of data offloaded to GPU in PS_2 and PS_4.

To model the obtained weak scaling results, we start from equation (39) for the execution time of a single iteration of the main loop. In weak scaling tests, the mesh size is increased proportionally to the number of MPI processes N_p, and therefore the execution time is given by

$$T_W(N_\mathrm{p}) = \alpha + \beta L + \gamma \frac{M N_\mathrm{p}}{N_\mathrm{p}} + \beta V \frac{M N_\mathrm{p}}{N_\mathrm{p}^2} = \alpha + \beta L + \gamma M + \beta V \frac{M}{N_\mathrm{p}} . \quad (42)$$

The weak scaling efficiency

$$E_W(N_\mathrm{p}) = \frac{T_W(1)}{T_W(N_\mathrm{p})} , \quad (43)$$

can thus be modeled by

$$E_W(N_\mathrm{p}) = \frac{1}{a + b/N_\mathrm{p}} . \quad (44)$$

We fitted this model to data presented in Figure 15 and the obtained fit parameters are given in Table 16. Note that the model is fitted to CUDA/MPI scaling data only for $N_\mathrm{p} \geq 4$. The agreement between the model and measurement data is very good, although we note that the parameter b has negative values, contrary to its expected relation to the communication cost. Therefore, we conclude that the above expression can be successfully used for modeling of the performance of all three MPI-based implementations.

To summarize, in all test results of MPI-based programs, we note an expected declining of the scaling efficiency. This is due to the introduction of distributed transposes of data, creating overhead that negatively impacts scaling efficiency. It is most evident in both strong and weak scaling tests of the CUDA/MPI version, as the transpose algorithm is inferior to the one provided by FFTW, used in OpenMP/MPI implementation. In our tests, all three MPI versions of programs failed to achieve actual speedup ($S(N_\mathrm{p}) > 1$) on less than $N_\mathrm{p} = 4$ nodes, due to the introduction of these transpose routines. We therefore recommend using MPI versions only on $N_\mathrm{p} = 4$ or more cluster nodes.

Table 16. Values of obtained weak scaling model fit parameters a and b, as well as the estimated fit errors for MPI-parallelized programs

Program	a	Δa	b	Δb
imag3d-mpi	2.56	0.06	-0.9	0.3
real3d-mpi	2.32	0.05	-1.3	0.2
imag3d-mpicuda	0.70	0.02	-0.1	0.2
real3d-mpicuda	0.68	0.02	-0.1	0.2
imag3d-mpihetero	1.32	0.03	-0.6	0.2
real3d-mpihetero	1.27	0.03	-0.6	0.2

2.4.5. Selecting the Optimal Algorithm

In previous sections we presented results of detailed performance tests and associated models for all developed programs. Here we provide general guidelines for obtaining the best performance from each implementation.

We note that the extensive testing performed shows that the best performance can be achieved by evenly distributing the workload among the MPI processes and OpenMP threads, and by using mesh sizes which are optimal for FFT. In particular, the single-node OpenMP programs have the best performance if the number of spatial points in all directions, controlled by parameters `Nx`, `Ny` and `Nz`, is divisible by the number of OpenMP threads used. Similarly, the OpenMP/MPI implementation achieves the best performance if `Nx` and `Ny` are divisible by a product of the number of MPI processes and the number of OpenMP threads used.

On the other hand, CUDA implementation works best if all the data for a given mesh size can fit into the GPU memory. In the case of a small mesh size, CUDA programs may not be able to saturate all Streaming Multiprocessors (SM) of a given GPU, and in this case the OpenMP programs may be considered first. For CUDA/MPI programs, the best performance is achieved if `Nx` and `Ny` are divisible by a product of the number of MPI processes and the number of SMs in the GPU used. Note that CUDA-based implementations write output files (e.g., density profiles) by transferring data from GPU memory to host memory, where a single thread writes to a file. Therefore, these implementations may not be suitable for simulations that require the output to be written frequently, after a small number of time propagation steps.

Best performance of hybrid implementations can be obtained by following the same guidelines. Furthermore, we recommend using these programs with the optimization methods described in section 2.4.1 to optimally divide the work between host and device. If manual tuning of offload parameters is required, we recommend that the guidelines above are followed for both CPU and GPU portions of the mesh. In the case of large disparity in the performance of host or device, hybrid versions will not provide the lowest execution times, and in such cases the pure CPU or GPU implementations with or without MPI could be better suited.

In addition to the guidelines for mesh sizes presented above, all programs benefit from the mesh size which is also optimal for FFT. The best FFT performance on CPU with FFTW library is obtained if `Nx`, `Ny` and `Nz` can be expressed as $2^a 3^b 5^c 7^d 11^e 13^f$, where e and f are either 0 or 1, and the other exponents are non-negative integer numbers [60]. Similarly, for cuFFT the best performance is achieved for transform sizes of the form $2^a 3^b 5^c 7^d$. In hybrid implementations, the same applies to the host and device portions of `Nx` and `Ny`, i.e., `cpuNx`, `cpuNy`, `gpuNx` and `gpuNy`.

2.5. Applications

The applicability and accuracy of the NLS equation for the description of BEC properties has been established in early studies addressing time–of–flight expansion and collective modes of the condensate [61, 1]. More recently, numerical algorithms [22, 23, 25, 26] for solving the NLS equation (1), have been used to investigate: nonlinear frequency shifts of the collective mode induced by a harmonic modulation of interaction [62]; onset and features of Faraday waves in BEC [63, 64, 65]; phase–separation dynamics of two–component BECs [66]; geometric resonances in a BEC with two– and three–body interactions [67]; non-equilibrium dynamics of a stirred BEC [68, 69, 70]; localization of a two–component BEC in a random potential [71]; vortex lattice melting in the presence of impurities [72]; classification of fractional vortices in a spinor BEC [73]; BEC dynamics in a dimple trap [74, 75, 76]; dynamics of localized impurity in BEC [77]; analysis of localized states in a multi–component BEC [78]; BEC dynamics induced by a sudden change of an optical lattice geometry [79]; BEC dynamics in a gravito–optical surface trap [80]; emergence of a Bose glass phase in a BEC in the presence of disorder [81]; vortex dynamics [82]; vortex generation in a stirred BEC [83]; dynamics of a self–bound droplet in the presence of an attractive two–body and a very small repulsive three–body interaction

[84, 85]. The dynamics of light bullets in a cubic–quintic nonlinear medium based on an extension of equation (12) has been published in reference [86]. As an illustration, in section 3 we present results of the study addressing the phase–separation dynamics of two–component BECs [66].

Furthermore, the codes [22, 23] have been extended to incorporate effects of the spin–orbit coupling which are currently intensely explored in the systems of cold atoms. In particular, the following subjects have been investigated: phase separation in a spin–orbit coupled BEC [87]; soliton dynamics in spin–orbit coupled BEC [88, 89, 90, 91]; dynamics of a BEC with spin–angular momentum coupling [92]. In section 4 we present recently obtained results about dipole mode and breathing mode excitations of a BEC in the presence of the coupling of pseudo spin and angular momentum [66].

The algorithms [28, 29, 30] for solving the dipolar NLS equation (11) have been applied to the study of: vortex dynamics in a dipolar BEC [93, 94, 95, 96]; phase separation in a binary dipolar BEC [97]; dynamics of a dipolar BEC droplet [98]; stability and dynamics of a dipolar Bose–Fermi mixture [99, 100]; solitons in a dipolar BEC [101, 102, 103, 104, 105, 106, 107]; self–trapping of a dipolar BEC in a double–well [108]. As a final illustration, in section 5 we present a study of vortex formation in a dipolar BEC.

3. Phase Separation Dynamics of a Two–Component BEC

Systems of multi–component BECs involve an additional degree of freedom and enrich the physics that can be explored within cold atom setups. These systems are either realized as mixtures of different bosonic atoms or different hyperfine states of the same atomic species. In this section we extend the previously described numerical algorithm and apply it for a study of a two–component system, described by two condensate wave functions $\psi_1(x,t)$ and $\psi_2(x,t)$ in a quasi one–dimensional geometry. The ground state configuration can be obtained as a minimum of the following energy functional

$$\begin{aligned} E_0 &= \int dx \left[(\psi_1^*(x)\ \psi_2^*(x))\, H_0 \begin{pmatrix} \psi_1(x) \\ \psi_2(x) \end{pmatrix} \right. \\ &+ \left. \frac{g_{11}}{2}|\psi_1(x)|^4 + \frac{g_{22}}{2}|\psi_2(x)|^4 + g_{12}|\psi_1(x)|^2|\psi_2(x)|^2 \right], \qquad (45) \end{aligned}$$

where H_0 corresponds to the non–interacting Hamiltonian

$$H_0 = \left(-\frac{1}{2}\frac{\partial^2}{\partial x^2} + V(x)\right)\mathcal{I}_2\,, \tag{46}$$

with $\mathcal{I}_2$ being a 2×2 identity matrix. As the total number of atoms of each species is conserved, the normalization condition is implemented as $\int dx|\psi_1(x)|^2 = \int dx|\psi_2(x)|^2 = 1$. The expression (45) encompasses three interaction constants: g_{11} for atoms of the first species, g_{22} for atoms of the second species and g_{12} interaction constant between the atoms of the two species. We assume that all these interactions are repulsive, i.e., $g_{11} \geq 0$, $g_{22} \geq 0$, $g_{12} \geq 0$. For $g_{12} = 0$, the two species do not "feel" each other and energy defined in (45) consists of two copies of single–component energy (5). In this case density distributions of the two species $|\psi_1(x)|^2$ and $|\psi_2(x)|^2$ may fully overlap in space. As intra–component repulsion g_{12} gets stronger, two–component mixtures may exhibit phase separation [13]. Simply, as it is energetically costly for the two species to overlap, they spatially separate instead. In the homogeneous case $V(x) = 0$, the necessary condition for the onset of phase separation is given by [109]

$$g_{11}g_{22} < g_{12}^2\,. \tag{47}$$

Often the two–component condensate is not prepared as the ground state of (45), but using a dynamical protocol [110, 111, 112] that we describe in the following. The dynamics of a two-component condensate in the mean–field regime is governed by two coupled Gross–Pitaevskii equations [113]:

$$i\frac{\partial\psi_1(x,t)}{\partial t} = \left[-\frac{1}{2}\frac{\partial^2}{x^2} + \frac{1}{2}\lambda^2x^2 + g_{11}|\psi_1(x,t)|^2 + g_{12}|\psi_2(x,t)|^2\right]\psi_1(x,t)\,, \tag{48}$$

$$i\frac{\partial\psi_2(x,t)}{\partial t} = \left[-\frac{1}{2}\frac{\partial^2}{x^2} + \frac{1}{2}\lambda^2x^2 + g_{12}|\psi_1(x,t)|^2 + g_{22}|\psi_2(x,t)|^2\right]\psi_2(x,t)\,. \tag{49}$$

Wave functions of the condensates $\psi_1(x,t)$ and $\psi_2(x,t)$ are normalized to unity and λ is the strength of the harmonic trap $V(x) = \lambda^2x^2/2$. The non–equilibrium dynamics of a two–component BEC captured by the coupled NLS equations (48) and (49) has attracted lot of theoretical interest [114, 115, 116, 117, 118].

Initially, a single-component condensate is produced. We model this situation as a two–component BEC with $g_{11} = g_{22} = g_{12}$ and prepare initial state as the ground state of (45) for $g_{11} = g_{22} = g_{12}$. At $t = 0$ a pulse is applied and half

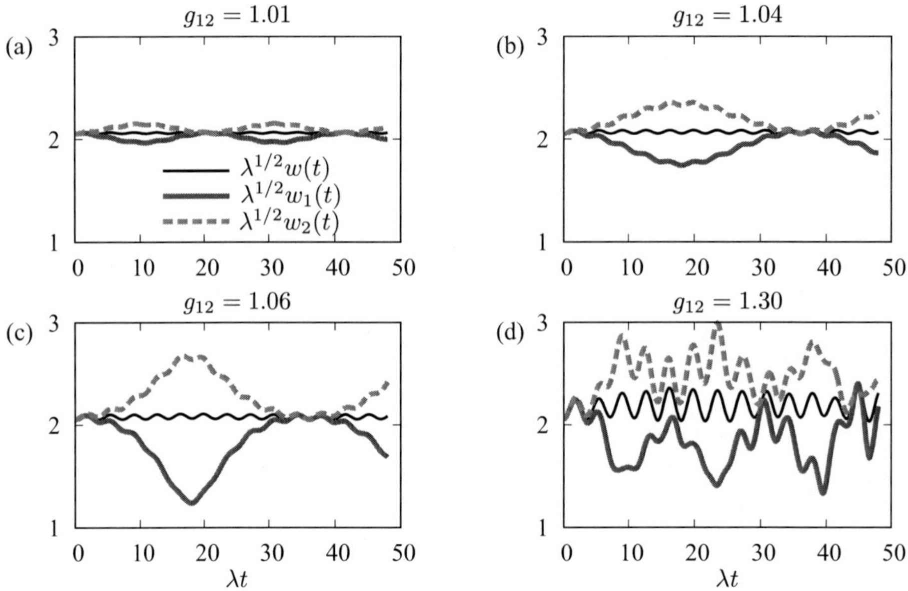

Figure 16. Time evolution of condensate widths for several values of g_{12}: (a) $g_{12} = 1.01$, (b) $g_{12} = 1.04$, (c) $g_{12} = 1.06$, (d) $g_{12} = 1.3$. Other parameters: $\lambda = 10^{-3}$, $g_{11} = 1$, $g_{22} = 1.01$.

the atoms are transfered into another hyperfine state. An important assumption of our analysis is that the process is completely spatially symmetric, i.e. the two BECs overlap perfectly at the initial moment. Due to the transfer, values of g_{22} and g_{12} are suddenly changed and dynamics of the two components is started. We monitor the time dependence of the widths of the two components,

$$w_{1,2}^2(t) = \int_{-\infty}^{\infty} dx\, x^2 |\psi_{1,2}(x,t)|^2 \,, \tag{50}$$

as well as the total width $w(t) = \sqrt{(w_1(t)^2 + w_2(t)^2)/2}$. Typical results obtained for different final values of g_{12} and $\lambda = 10^{-3}$, $g_{11} = 1$, $g_{22} = 1.01$ are presented in Figure 16. We observe fast oscillations in the total width with a frequency unaffected by a change in g_{12}. These oscillations correspond to the so–called breathing mode. In a quasi one–dimensional geometry it turns out that their period is $2\pi/(\sqrt{3}\lambda)$. On top of this, we notice slow oscillations in the width of each of the two component for $g_{12} = 1.01$ as shown in Figure 16(a).

These out–of–phase oscillations correspond to the "spin" mode, and their frequency decreases with increase in g_{12}. In particular, the oscillation period of the spin mode is about 20 for $g_{12} = 1.01$ (see Figure 16(a)) and about 30 for $g_{12} = 1.04$ (see Figure 16(b)). Above a certain threshold value of g_{12}, oscillation amplitudes sharply increase and the width dynamics becomes aperiodic and irregular (see Figure 16(c) for $g_{12} = 1.06$). Irregularity of the width dynamics is even more apparent for $g_{12} = 1.3$ as shown in Figure 16(d).

In Figure 17 we show the dynamics of the "spin density" $|\psi_1(x,t)|^2 - |\psi_2(x,t)|^2$ for the same parameters as in Figure 16. The plots Figure 17(a) and (b) show regular oscillations. By comparing plots Figure 17(b) and (c) we note a sharp change in the color scales that matches the sharp increase in the oscillation amplitude observed earlier. Finally, in Figure 17(d) we observe a clear onset of pattern dynamics. By applying a suitable linear stability analysis, it was shown that the onset of aperiodic motion is marked by emergence of an unstable mode around a referent stationary state [66]. This behavior is known as modulation instability and by investigating a range of values $\lambda \in \left(10^{-5}, 10^{-1}\right)$, we found that in a realistic experimental situation, stronger trapping potential strength λ shifts the instability condition (47) toward higher values of g_{12}.

4. Excitations of a BEC with Spin-Angular Momentum Coupling

Experimental realization of an effective spin–orbit coupling in ultracold atom systems [119, 120, 121, 122, 123, 124, 125] has allowed for new quantum phases to be explored. Bosonic systems with spin–orbit coupling are interesting as they have no direct analogues in condensed matter systems and provide a new research playground [126, 127]. In recent references [128, 129, 130, 131] the following Hamiltonian for a two–component bosonic system has been introduced:

$$H_{\mathrm{SOC}} = \left(-\frac{1}{2}\Delta + \frac{r^2}{2}\right)\mathcal{I}_2 + \frac{\Omega^2 r^2}{2}\begin{pmatrix} 1 & e^{-2i\phi} \\ e^{2i\phi} & 1 \end{pmatrix}, \qquad (51)$$

where $\mathcal{I}_2$ is a 2×2 identity matrix and the effective spin $1/2$ comes from the two bosonic components involved. The system is assumed to be effectively two–dimensional (tightly trapped in the longitudinal direction). The value of Ω is set by the intensity of two co–propagating Laguerre–Gauss laser beams

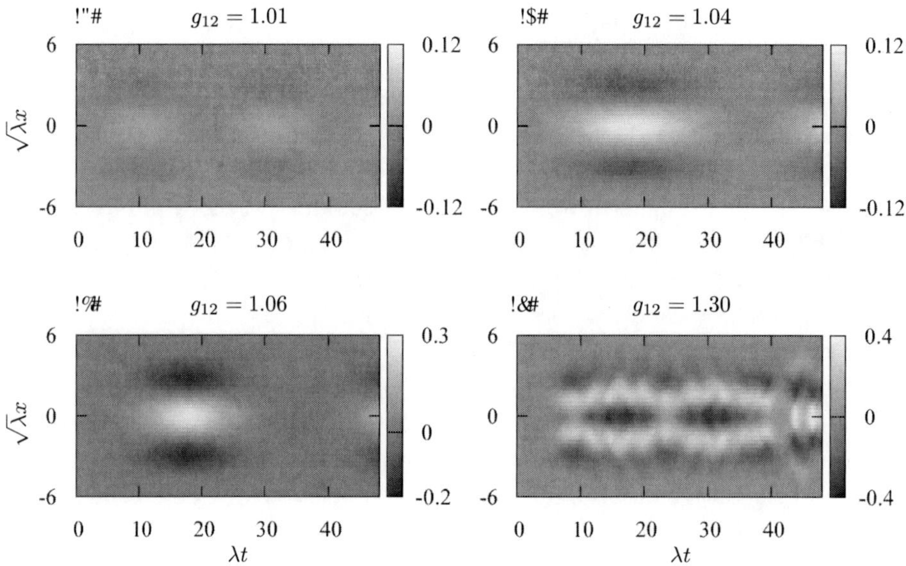

Figure 17. Spin dynamics illustrated via the density difference given by $\lambda^{-1/2}\left(|\psi_1(x,t)|^2 - |\psi_2(x,t)|^2\right)$, induced by a quench of g_{12} and g_{22}. Parameters: (a) $g_{12} = 1.01$, (b) $g_{12} = 1.04$, (c) $g_{12} = 1.06$, (d) $g_{12} = 1.3$ and $\lambda = 10^{-3}$, $g_{11} = 1$, $g_{22} = 1.01$.

that carry a unit of angular momentum in the opposite rotational directions. The last, ϕ–dependent term, where ϕ is the polar angle, provides the coupling between the spin and angular momentum, as can be explicated by using a proper unitary transformation [129]. In this section we extend the algorithms presented in subsection 2.2 and use them to investigate excitations of a BEC with spin-angular momentum coupling as defined in equation (51).

The total energy per particle of the condensed state with the order parameter $(\psi_1(\mathbf{r})\ \psi_2(\mathbf{r}))^T$ is given by

$$E_0 = \int d\mathbf{r}\left[(\psi_1^*\ \psi_2^*)\, H_{\mathrm{SOC}}\begin{pmatrix}\psi_1\\ \psi_2\end{pmatrix} + \frac{g}{2}|\psi_1|^4 + \frac{g}{2}|\psi_2|^4 + g|\psi_1|^2|\psi_2|^2\right], \quad (52)$$

where we assume spin–symmetric interactions $g_{11} = g_{22} = g_{12} = g$. Due to off–diagonal terms in (51), which allow for the conversion of one atomic state into the other, only the total number of atoms is conserved and the normalization condition is given by $\int d\mathbf{r}\left(|\psi_1(\mathbf{r})|^{\mathbf{2}} + |\psi_{\mathbf{2}}(\mathbf{r})|^{\mathbf{2}}\right) = 1$.

To implement the spin–angular momentum coupling in the propagation routine, we split the step of imaginary–time propagation into four pieces. As usual, in the first step we take into account harmonic potential and nonlinear terms, according to equation (13). The second step comprises the spin–angular momentum coupling, and in the last two steps we perform integrations of the type (14) with respect to the two spatial variables x and y. Explicitly, the off–diagonal terms from (51) are treated separately starting from

$$-\frac{\partial\psi_1(\mathbf{r},\tau)}{\partial\tau} = e^{-2i\phi}\,c(r)\psi_2(\mathbf{r},\tau), \tag{53}$$

$$-\frac{\partial\psi_2(\mathbf{r},\tau)}{\partial\tau} = e^{2i\phi}\,c(r)\psi_1(\mathbf{r},\tau), \tag{54}$$

where $c(r) = \frac{1}{2}\Omega^2 r^2$. By differentiating the two last equations we obtain two independent equations

$$\frac{\partial^2\psi_1(\mathbf{r},\tau)}{\partial\tau^2} = c(r)^2\psi_1(\mathbf{r},\tau), \quad \frac{\partial^2\psi_2(\mathbf{r},\tau)}{\partial\tau^2} = c(r)^2\psi_2(\mathbf{r},\tau) \tag{55}$$

that are solved by

$$\begin{aligned}\psi_1(\mathbf{r},\tau) &= A_1(\mathbf{r})\,e^{-c(r)\tau} + B_1(\mathbf{r})\,e^{c(r)\tau}\,,\\ \psi_2(\mathbf{r},\tau) &= A_2(\mathbf{r})\,e^{-c(r)\tau} + B_2(\mathbf{r})\,e^{c(r)\tau}\,.\end{aligned} \tag{56}$$

From the initial conditions for the second part of the n–th integration step at $\tau' = \left(n+\frac{1}{4}\right)\varepsilon$ we obtain a set of algebraic equations for the coefficients

$$\begin{aligned}A_1(\mathbf{r}) + B_1(\mathbf{r}) &= \psi_1(\mathbf{r},\tau')\,,\\ -c(r)A_1(\mathbf{r}) + c(r)B_1(\mathbf{r}) &= -e^{-2i\phi}c(r)\,\psi_2(\mathbf{r},\tau')\,,\\ A_2(\mathbf{r}) + B_2(\mathbf{r}) &= \psi_2(\mathbf{r},\tau')\,,\\ -c(r)A_2(\mathbf{r}) + c(r)B_2(\mathbf{r}) &= -e^{2i\phi}\,c(r)\psi_1(\mathbf{r},\tau')\,,\end{aligned}$$

that finally leads to

$$\begin{aligned}A_1(\mathbf{r}) &= \frac{1}{2}\left[\psi_1(\mathbf{r},\tau') + e^{-2i\phi}\,\psi_2(\mathbf{r},\tau')\right]\,,\\ B_1(\mathbf{r}) &= \frac{1}{2}\left[\psi_1(\mathbf{r},\tau') - e^{-2i\phi}\,\psi_2(\mathbf{r},\tau')\right]\,,\\ A_2(\mathbf{r}) &= \frac{1}{2}\left[\psi_2(\mathbf{r},\tau') + e^{2i\phi}\,\psi_1(\mathbf{r},\tau')\right]\,,\\ B_2(\mathbf{r}) &= \frac{1}{2}\left[\psi_2(\mathbf{r},\tau') - e^{2i\phi}\,\psi_1(\mathbf{r},\tau')\right]\,.\end{aligned} \tag{57}$$

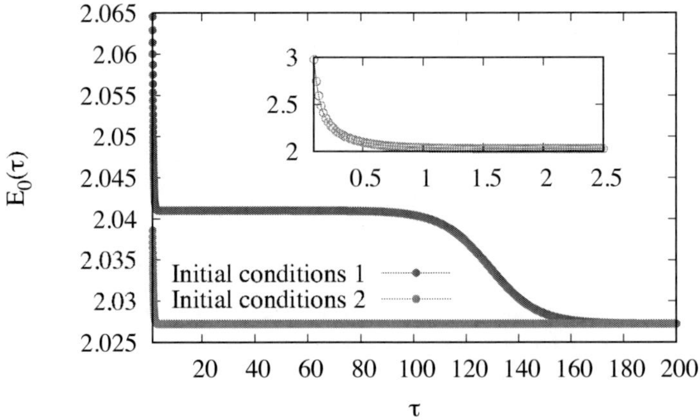

Figure 18. The energy of the ground state $E_0(\tau)$ defined in equation (52) vs. imaginary time τ obtained for different initial conditions. In the first case, we start from a random initial state and observe slow convergence toward the global minimum. In the second case, initial conditions have proper dependence on the polar angle ϕ, as given in equation (58) and the convergence toward the ground state is exponentially fast as shown in the inset. Parameters: $\Omega = 3.2$ and $g = 1$.

In our numerical implementation we make direct use of equations (56) and (57).

Before turning to numerical simulations, we first briefly present an important analytical result available for $g = 0$. General symmetry considerations imply that the ground state configurations of (52) for $g = 0$ can be expressed in a form

$$\phi_m(r, \phi) = \frac{e^{im\phi}}{\sqrt{2\pi}} \begin{pmatrix} f_m(r)e^{-i\phi} \\ g_m(r)e^{i\phi} \end{pmatrix}, \tag{58}$$

where m is an integer. In particular for $\Omega < \Omega_c$, with $\Omega_c = 3.35$ there are two degenerate ground states $m = \pm 1$ and for $\Omega > \Omega_c$ the ground state configuration corresponds to $m = 0$. The $m = \pm 1$ states are denoted as half–vortices, and $m = 0$ state is named vortex–antivortex pair. In order to explore ground state configurations for a finite value of interaction constant g, we perform minimization of energy functional (52) using imaginary time propagation. How quickly we reach the ground state strongly depends on the choice of initial configuration as shown in Figure 18. Starting with random initial values, we find that propagation routine may first reach a metastable configuration (local minimum of E_0) and eventually, only after very long propagation time, it selects the right minimum. With a better choice of initial configuration, motivated by exact results (58), the right minimum is found with substantially less iterations. Considerations of this kind are relevant whenever there are several configurations that are nearly degenerate – i.e., have close values of the corresponding energies (52) and represent local minima of the functional (52).

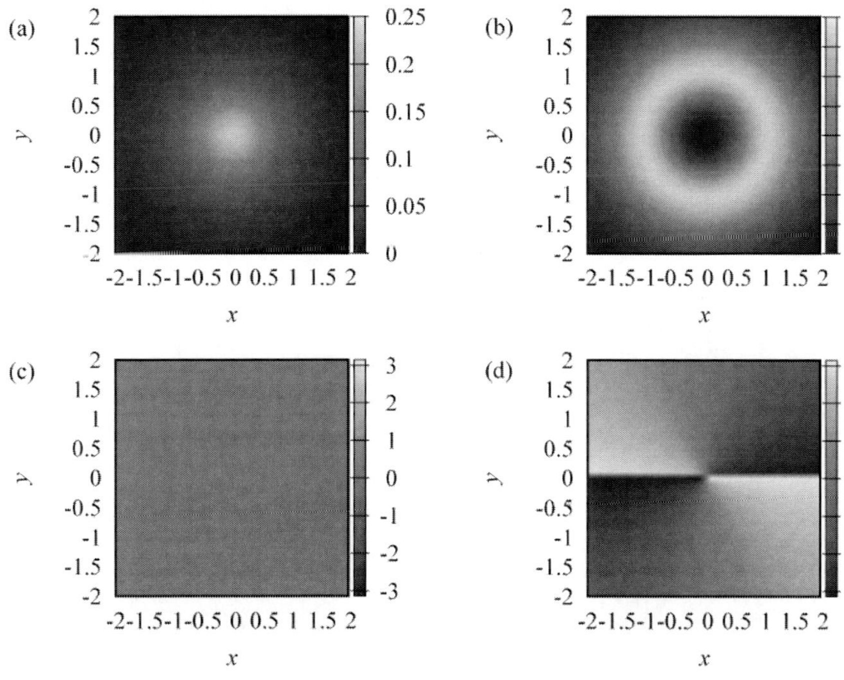

Figure 19. The $m = 1$ ground state obtained for $\Omega = 3.2$, $g = 1$. We plot: (a) the density distribution of the first component $|\psi_1(\mathbf{r})|^2$, (b) the density distribution of the second component $|\psi_2(\mathbf{r})|^2$, (c) the argument of $\psi_1(\mathbf{r})$ and (d) the argument of $\psi_2(\mathbf{r})$.

Two examples of the ground state configurations are given in Figs. 19 and 20. We find that at finite values of g, ground states can also be distinguished according to their dependence on polar angle ϕ, as given in equation (58). In Figure 19 we present the ground state solution for $\Omega = 3.2$ and $g = 1$. This is a half–skyrmion configuration characterized by $m = 1$. We note that the first component is confined around the trap center and its phase dependence is trivial. In contrast, the second bosonic species exhibits double phase winding, see Figure 19(d). In Figure 20 we present the ground state solution for $\Omega = 3.5$ and $g = 1$. The densities of the two components overlap in space, Figure 20(a) and (b), and exhibit phase winding in opposite directions, Figure 20 (c) and (d). This is an example of $m = 0$ state called a vortex–antivortex pair.

In general, quantum phases can be distinguished according to their low–lying spectra. In cold atom experiments routinely accessible with great precision are breathing mode and dipole mode excitations. To probe these modes, initially the system is prepared in its ground state and then a sudden quench of the potential $V(\mathbf{r})$ is performed. We now theoretically address these excitations for the half-skyrmion and vortex–antivortex state.

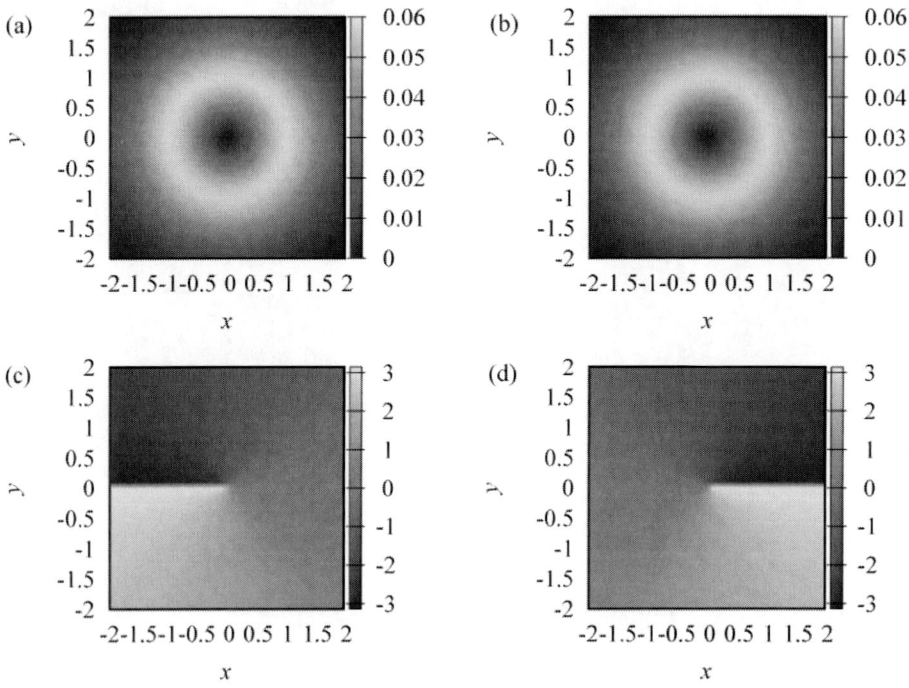

Figure 20. The $m = 0$ ground state obtained for $\Omega = 3.5$, $g = 1$. We plot: (a) the density distribution of the first component $|\psi_1(\mathbf{r})|^2$, (b) the density distribution of the second component $|\psi_2(\mathbf{r})|^2$, (c) the argument of $\psi_1(\mathbf{r})$ and (d) the argument of $\psi_2(\mathbf{r})$.

The dynamics of a BEC with spin–angular momentum coupling is governed by two coupled time–dependent Gross-Pitaevskii equations for $\psi_1(\mathbf{r}, t)$ and $\psi_2(\mathbf{r}, t)$:

$$i\frac{\partial \psi_1}{\partial t} = \left[-\frac{1}{2}\left(\frac{\partial^2}{\partial x^2} + \frac{\partial^2}{\partial y^2}\right) + V(r) + \frac{r^2\Omega^2}{2} + g\left(|\psi_1|^2 + |\psi_2|^2\right)\right]\psi_1 + \frac{r^2\Omega^2}{2}e^{-2i\phi}\psi_2 \,, \quad (59)$$

$$i\frac{\partial \psi_2}{\partial t} = \left[-\frac{1}{2}\left(\frac{\partial^2}{\partial x^2} + \frac{\partial^2}{\partial y^2}\right) + V(r) + \frac{r^2\Omega^2}{2} + g\left(|\psi_2|^2 + |\psi_1|^2\right)\right]\psi_2 + \frac{r^2\Omega^2}{2}e^{2i\phi}\psi_1 \,. \quad (60)$$

The trapping potential is assumed to be harmonic, initially set to $V(r) = r^2/2$. To induce a breathing mode, we perturb the trap strength

$$V_{\mathrm{pert}} = (1+\eta)\frac{r^2}{2}, \quad (61)$$

and calculate the time evolution of the width of the probability distribution,

$$\langle r^2(t)\rangle = \int_0^{2\pi} d\phi \int_0^{\infty} dr\, r^3 \left[|\psi_1(\mathbf{r}, t)|^2 + |\psi_2(\mathbf{r}, t)|^2\right] . \quad (62)$$

By inspecting the corresponding Fourier transform, we identify the lowest excited frequency as the breathing mode ω_B.

In a typical harmonically trapped two–dimensional system, without the spin–orbit coupling $\Omega = 0$, the Gross–Pitaevskii description predicts that the breathing mode frequency is $\omega_B = 2$ (in the units of the harmonic trap frequency), independent of interaction constant g [132]. Analytical arguments and numerical results (not presented here) show that this general conclusion holds true for the vortex–antivortex state. On the other hand, the breathing mode frequency of a half–skyrmion state depends on both the coupling Ω and on the interaction strength g. Illustrative results for $\langle r^2(t)\rangle$ are presented in Figure 21(a) for $\Omega = 3.2$, $\eta = 0.1$ and several values of g. From the related Fourier transforms shown in Figure 21(b), we find that the breathing mode frequency increases for several percent with g.

To excite a dipole mode, we consider a shift of the trap bottom in x direction,

$$V_{\mathrm{pert}} = \frac{r^2}{2} - \delta x \frac{r}{2}\left(e^{i\phi} + e^{-i\phi}\right), \quad (63)$$

and monitor the motion of the center of mass of the system in that direction,

$$\langle x(t)\rangle = \int_0^{2\pi} d\phi \frac{e^{i\phi} + e^{-i\phi}}{2} \int_0^{\infty} dr\, r^2 \left[|\psi_1(\mathbf{r}, t)|^2 + |\psi_2(\mathbf{r}, t)|^2\right] , \quad (64)$$

as well as $\langle y(t)\rangle$.

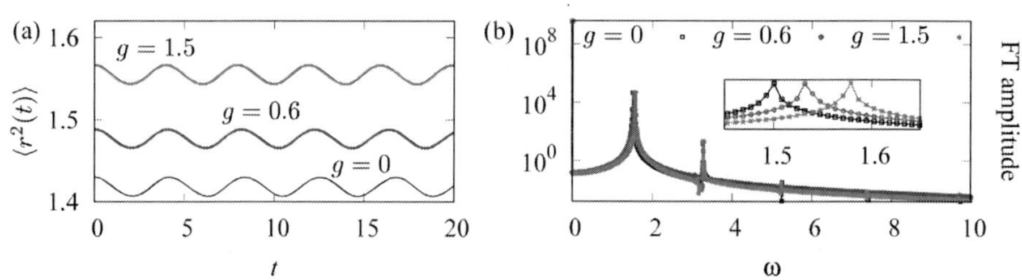

Figure 21. Breathing mode oscillations in half–skyrmion phase: (a) $\langle r^2(t) \rangle$ versus t and (b) corresponding Fourier transform. From the inset we observe increase of the breathing mode frequency with g. Motion is induced by changing harmonic trap potential as $\frac{r^2}{2} \rightarrow 1.01\frac{r^2}{2}$, $\Omega = 3.2$.

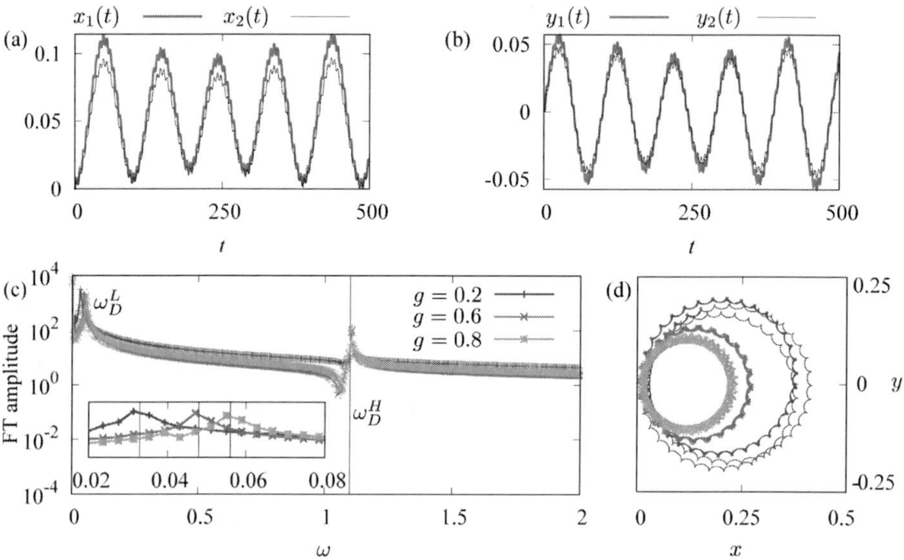

Figure 22. Dipole mode oscillations of half–skyrmion state in interacting case for $\Omega = 3.2$. Motion is induced by shifting the harmonic trap bottom for $\delta x = 0.01$. In (a) and (b) $g = 1$. In (c) motion of the center of mass, $y(t)$ versus $x(t)$, is plotted. The trajectory radius gets smaller with increasing g. In (d) vertical lines give results for ω_D^L (in the inset) and ω_D^H (in the main panel) obtained using the Bogoliubov method, and dots represent Fourier transform of $x(t)$.

In the conventional harmonically trapped system ($\Omega = 0$), this perturbation excites the Kohn mode – an oscillation with the trap frequency along x

axis unaffected by interactions. Illustrative results for a half–skyrmion state are presented in Figure 22 for $\Omega = 3.2$ and they obviously exhibit more complex behavior in comparison to the results for $\Omega = 0$. We observe that oscillatory motion in x direction is coupled to the motion along y direction. Moreover, we find that there are two frequencies involved. We label them as ω_D^L (low) and ω_D^H (high). In Figure 22(a) and 22(b) we see that two bosonic components oscillate in–phase in both directions. The frequency ω_D^L exhibits a strong increase with g, as is depicted in Figure 22(c). As the frequency ω_D^L gets larger, the induced oscillation amplitude gets weaker, Figure 22(d). For the considered case, the frequency ω_D^H is found to be almost independent of g, see Figure 22(c). Results based on the complementary Bogoliubov approach, match quite well to the numerical data obtained from direct numerical simulations of equations (59)–(60), see Figure 22(c).

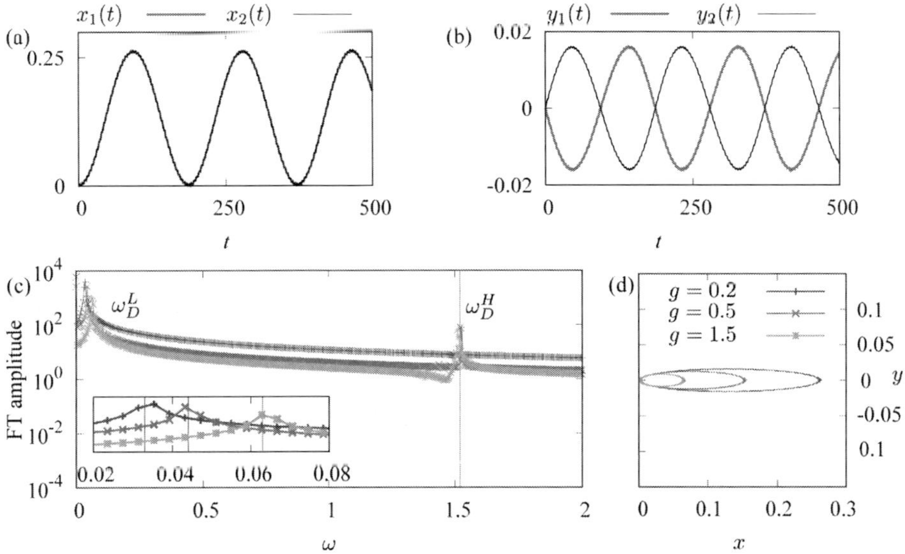

Figure 23. Dipole mode oscillations of the $m = 0$ solution in interacting case for $\Omega = 3.5$. Motion is induced by shifting harmonic trap bottom for $\delta x = 0.01$. In (a) and (b) $g = 0.2$. In (c) trajectory of the center of mass of a single bosonic component, $y_1(t)$ versus $x_1(t)$, is plotted. In (d) vertical lines give results for ω_D^L (in the inset) and ω_D^H (in the main panel) obtained using Bogoliubov method and dots represent Fourier transform of $x(t)$.

In the vortex–antivortex state, the two bosonic components exhibit an out–of–phase oscillation in y direction, see Figure 23(b), and consequently the center of mass only oscillates in x direction with the frequency ω_D^L that exhibits an increase with g, see Figure 23(c). The trajectory of the center of mass of each of the components is given by an ellipse, which is strongly elongated in x direction, see Figure 23(d).

5. Study of Critical Velocity for Vortex Formation

As we have already seen, the programs presented in this chapter can be used to model and study a variety of systems. In this section we demonstrate the versatility of developed programs on a simulation of vortex formation in a BEC. We use the MPI-based programs to simulate the effects of a moving obstacle in an oblate atomic BEC. The obstacle, a repulsive Gaussian laser beam, moves through the condensate and sheds quantum vortices, elementary excitation of a superfluid [14, 13]. The vortices appear only when the obstacle is moving above some critical speed, consistent with the Landau's criterion of superfluidity. At low obstacle velocities above a critical value, vortex dipoles emerge, and as the speed is increased further, individual vortices and rotating vortex pairs are also formed.

In the first subsection we document how our programs can be used to simulate the experiment reported in reference [133], and compare experimental and numerical results. The experiment measured critical velocity for the emergence of vortex dipoles and rotating vortex pairs in a BEC of sodium atoms (^{23}Na), which do not exhibit the dipolar interaction. Therefore, in this section we also demonstrate how our programs can be modified to switch off the dipolar interaction term when solving NLS equation. The agreement of the results obtained numerically and experimental observations provides an external check of the correctness of our algorithms and their implementations.

In the second subsection we investigate the formation of vortices in a dipolar BEC of dysprosium atoms (^{164}Dy). Our simulations follow the same methodology as the experiment with sodium atoms, only with atomic species exhibiting strong dipolar interaction. We study effects of the dipolar interaction on the critical velocity for the emergence of vortices, as well as the interplay between contact and dipolar interaction. The visualization extensions presented in reference [42] have proven to be indispensable during these simulations, as they allow much easier study of the results and control of the simulation.

5.1. Formation of Vortices in a BEC

BEC is a superfluid quantum liquid and one of its hallmarks are quantized vortices, which appear as elementary excitations of the system. Quantization of vortices is connected to excitation spectrum and, according to Landau's criterion, leads to the existence of a minimal velocity an obstacle moving through the superfluid has to have in order to generate such elementary excitations. This critical velocity v_c can be experimentally measured and, in principle, depends on the experimental protocol used.

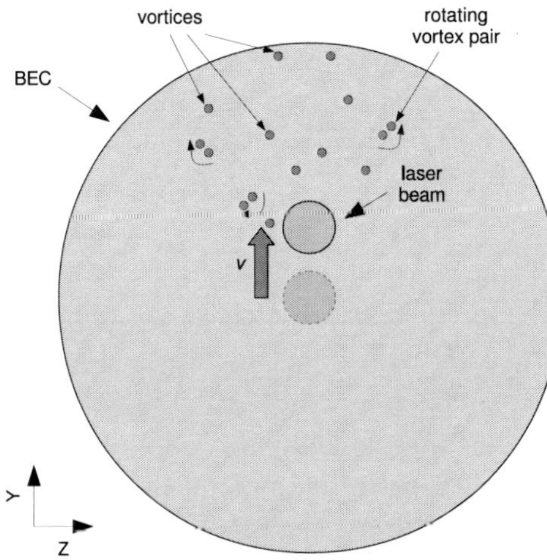

Figure 24. Illustration of the experimental setup used in reference [133] to study vortex formation in a BEC of sodium atoms. A repulsive Gaussian laser beam is initially at the center of the condensate and moves along y direction with a constant velocity v until it reaches its final position. The beam is then switched off during a period of 0.5 s. For sufficiently large velocity vortices and vortex pairs are generated.

Reference [133] reports the study of vortex formation in an oblate BEC of $N_{\text{at}} = 3.2 \times 10^6$ sodium atoms. The trapping potential frequencies used are $(\omega_x, \omega_y, \omega_z) = 2\pi \times (9,\ 9,\ 400)$ Hz, and the s-wave scattering length determining the strength of contact interaction was $a_s = 51.9\, a_0$, where $a_0 = 0.0529$ nm is the Bohr radius. The moving obstacle is realized by a repulsive Gaussian laser beam corresponding to an additional potential of the form

$$V_{\rm B}(\mathbf{r};t) = V_0 \exp\left\{-2\frac{[y-y_0(t)]^2+z^2}{\sigma^2}\right\}, \qquad (65)$$

where V_0 represents the strength of the beam, σ is the $1/e^2$ beam waist, while $y_0(t) = vt$ determines the center of the beam position $(0, y_0(t), 0)$. In our simulations, as in the experiment, the initial position of the beam is at the condensate center and then it moves with the velocity v along y direction for 24 μm. Afterwards, the beam is switched off linearly during 0.5 s. Illustration of the experimental setup is given in Figure 24. Using imaginary-time propagation, we calculate the ground state of the system with the trapping potential $V(\mathbf{r}) + V_{\rm B}(\mathbf{r};t)$, where $V(\mathbf{r})$ corresponds to the harmonic part, with the frequencies given above. The experiment measured critical velocity for the emergence of vortex dipoles, pairs of vortices of the opposite sign. We numerically addressed this setup and calculated this critical velocity for the values of the parameters $\sigma = 1.3\,l$ and $V_0 = 250\,\hbar\tilde{\omega}$, where $l = 6.988\,\mu$m is the harmonic oscillator length for the referent frequency $\tilde{\omega} = \omega_x$.

Before runnnig simulations with the parameters described above, we had to implement the time-dependent potential $V(\mathbf{r}) + V_{\rm B}(\mathbf{r};t)$. Initialization of $V(\mathbf{r})$ remains the same, i.e., it is initialized before the main loop and stored in an array, while calculation of $V_{\rm B}(\mathbf{r};t)$ is implemented in a separate function that is called in each iteration within the main loop. Two distinct phases of $V_{\rm B}(\mathbf{r};t)$ exist: one relating to the movement of the beam, and the other relating to the beam shutdown during 0.5 s. Each phase is implemented as a separate function, and the active phase is determined by the current time, i.e., iteration number.

Figure 25 shows several snapshots of a typical dynamical evolution of the system. Imaginary-time propagation yields ground state of the system, which is shown in Figure 25(a), where we plot integrated 2D density profile in y-z plane. To model local inhomogeneities always present in the experiment, we add uniformly distributed random noise to the wave function of the order of 10%. Such modified ground state represents initial state of the system, and the beam starts to move along y direction at time $t = 0$ with the speed $v =$ 1.26 mm/s. Figures 25(b) and 25(c) show 2D density profile of the system at times $t = 52.17$ ms and $t = 358.1$ ms, respectively. In Figure 25(b) the beam already reached its final position and its switch-off started. We can observe that several vortices (vortex dipoles) are generated and that the used speed exceeds the critical velocity. Generated vortices are stable and can be seen after extended

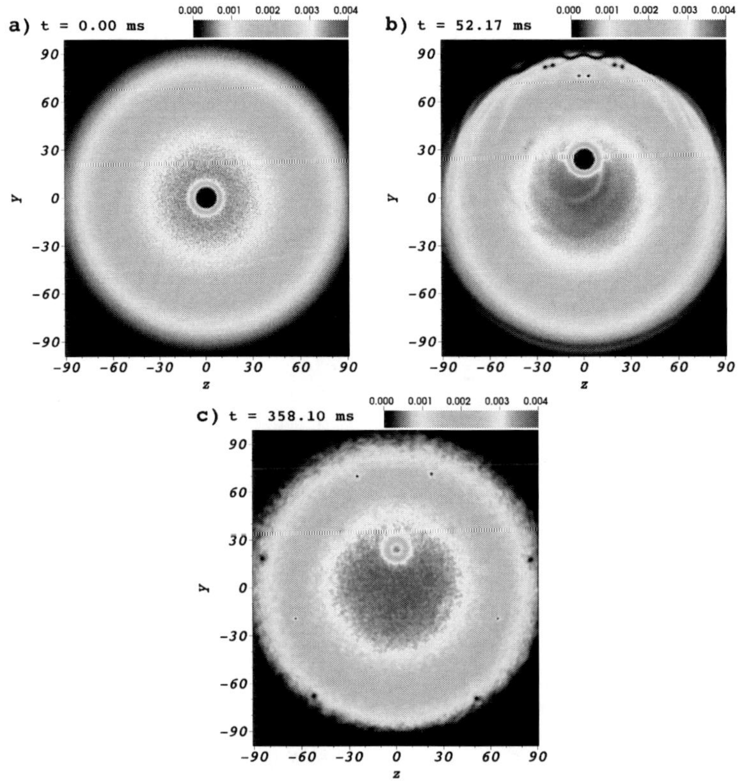

Figure 25. Time evolution of 2D density profile of BEC of sodium atoms with an obstacle moving at speed $v = 1.26$ mm/s $> v_c$, when several vortices are generated. Each panel shows integrated 2D density profile in y-z plane at different time t. All lengths are expressed in units of μm, and particle density is given in units N_{at}/l^2.

period of time, Figure 25(c). To confirm that we indeed have vortices and not just localized density minima, we plot the $x = 0$ slice of the phase of the wave function in Figure 26. We observe characteristic braiding and jumps of the phase in the vicinity of density minima, which is a well-known hallmark of a vortex.

When the obstacle is moving with an under-critical velocity, no vortices are generated, as in Figure 27. The velocity $v = 0.87$ mm/s is just slightly lower than the the critical one, and we can see precursors of vortices at the beam edge in Figures 27(a) and 27(b). At higher velocities, vortices emerge, as we have seen in Figure 25. At even higher velocities, Figure 28, rotating vortex pairs are generated, in addition to individual vortices.

The obtained results are in good agreement with experimental findings of reference [133] and show that the programs developed within this thesis can be successfully used to model BEC systems with contact interaction, even for the most complex setup, when vortices are generated due to a moving obstacle.

5.2. Effects of Dipolar Interactions on Vortex Formation in a BEC

To further test our programs, we model vortex formation in a BEC of dysprosium atoms (^{164}Dy), which exhibit strong magnetic dipole moment. In par-

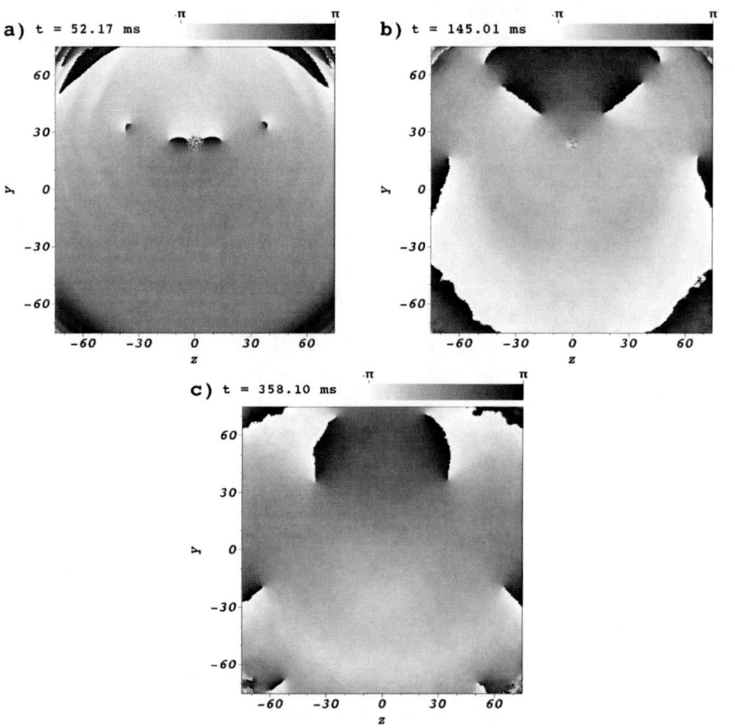

Figure 26. Time evolution of the wave function phase of BEC of sodium atoms with an obstacle moving at speed $v = 1.26$ mm/s $> v_c$, when several vortices are generated. Each panel shows $x = 0$ slice of the phase of the wave function in y-z plane at different time t. All lengths are expressed in units of μm.

ticular, we study effects of the dipolar interaction strength for varying s-wave scattering lengths on the critical velocity for the emergence of vortices, which was not investigated experimentally. Dysprosium atoms have the largest magnetic dipole moment ($m = 10\mu_B$) available in ultracold atom experiments, corresponding the characteristic dipole-dipole interaction length $a_{dd} = 132\,a_0$. External magnetic field can be used to align all atomic dipoles in the same direction, as well as to tune their strength up to a maximal value given above. The same applies to the s-wave scattering length, which determines the contact

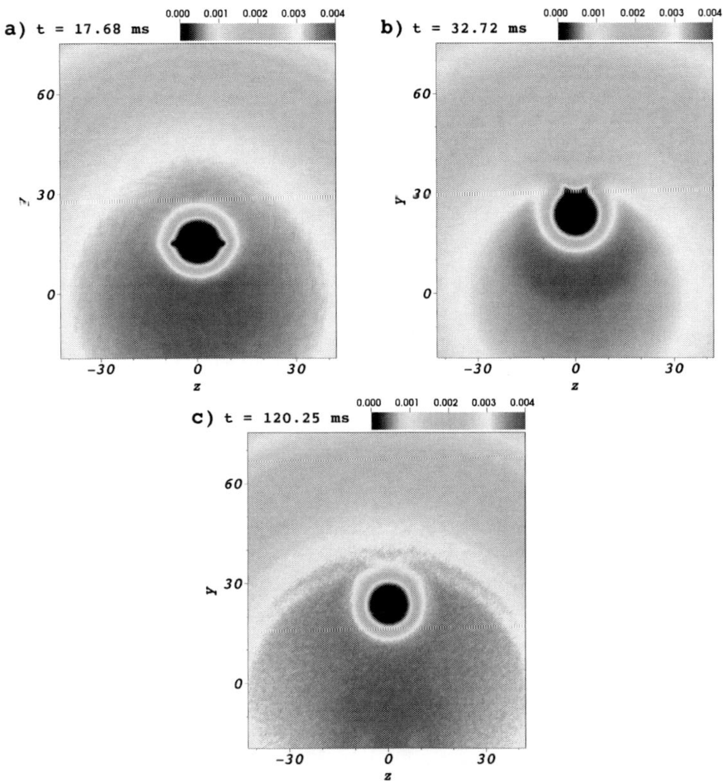

Figure 27. Time evolution of 2D density profile of BEC of sodium atoms with an obstacle moving at speed $v = 0.87$ mm/s $\lesssim v_c$, when no vortices are generated. Each panel shows integrated 2D density profile in y-z plane at different time t. All lengths are expressed in units of μm, and particle density is given in units N_{at}/l^2.

interaction strength and which can also be tuned using the Feshbach resonance technique. Following reference [134], we take the same range of possible values for a_s as for $a_{\rm dd}$, i.e., up to a maximal value of $132\,a_0$. With this we show how the developed programs can be used to model and theoretically address new physical phenomena, before they are studied experimentally. Such approach is essential for the design of many upcoming experiments, since otherwise it would be extremely difficult to predict what would be the relevant range of physical quantities to be measured, thus making it very challenging to perform

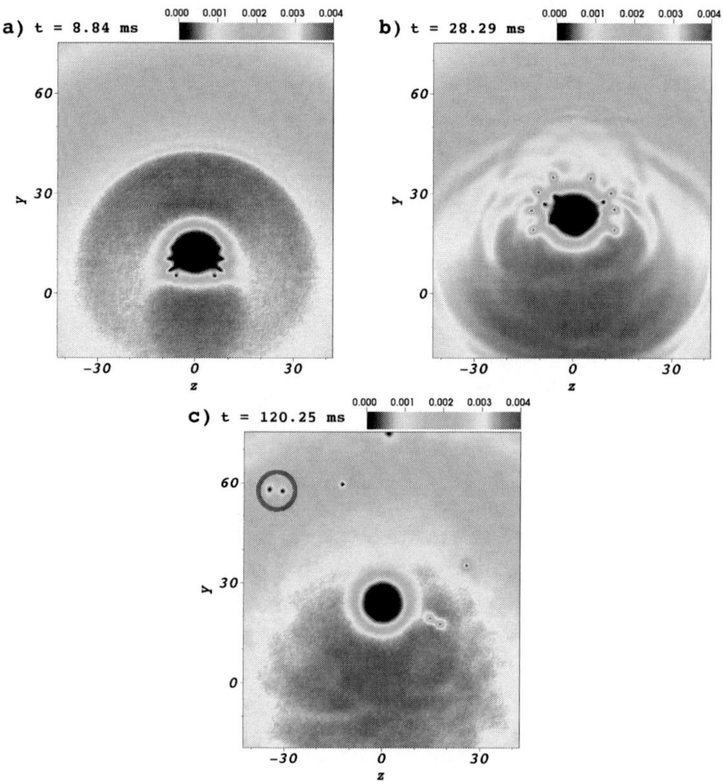

Figure 28. Time evolution of 2D density profile of BEC of sodium atoms with an obstacle moving at speed $v = 1.4$ mm/s $\gg v_c$, when rotating vortex pairs are also generated. Each panel shows integrated 2D density profile in y-z plane at different time t. A rotating vortex pair is highlighted in panel (c). All lengths are expressed in units of μm, and particle density is given in units $N_{\rm at}/l^2$.

the experiments. Having the results of detailed numerical simulations for the particular system enables experiments to target appropriate range of all relevant quantities, and to focus on discovering new phenomena.

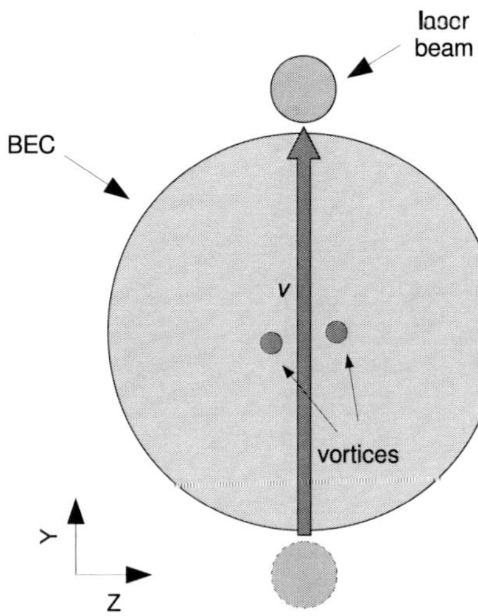

Figure 29. Illustration of the setup we use to study vortex formation in a dipolar BEC of dysprosium atoms. A repulsive Gaussian laser beam is initially outside of the condensate and moves along y direction with a constant velocity v until it reaches its final position on the other side. For sufficiently large velocity vortices are generated.

The experiment follows the same methodology as in previous section and uses the same parameters where applicable. Namely, the trap frequencies remain the same, i.e., $(\omega_x, \omega_y, \omega_z) = 2\pi \times (9, 9, 400)$ Hz, as well as strength of the beam $V_0 = 250\,\hbar\tilde{\omega}$. Due to the strong dipolar interactions affecting the stability of the BEC, we had to use a much smaller number of atoms, $N_{\rm at} = 8 \times 10^4$. This resulted in a much smaller BEC, so we had to reduce the $1/e^2$ beam waist to $\sigma = l$, where $l = 2.617\,\mu$m is harmonic oscillator length corresponding to dysprosium atoms. Also, we positioned the beam outside of the condensate, and move it all the way to the other side, as illustrated in Figure 29. The potential $V_{\rm B}(\mathbf{r};t)$ has the same form (65) as in previous section, just the center of the beam is now given by $y_0(t) = y_{00} + vt$, where $y_{00} = -15$ in units of l.

Modifications to the programs presented in the previous section can be reused for this numerical experiment and we only need to re-enable previously disabled calculation of the dipolar term in the main time propagation loop. As before, the ground state of the system is obtained using imaginary-time program, which then serves as the initial state of the real-time propagation program, with uniformly distributed random noise of 10% added. To calculate critical velocity v_c for the emergence of vortices, we search for the minimal speed of the laser beam for which two vortices appear, as illustrated in Figure 29.

Figure 30 shows several 2D density profiles of a typical dynamical evolu-

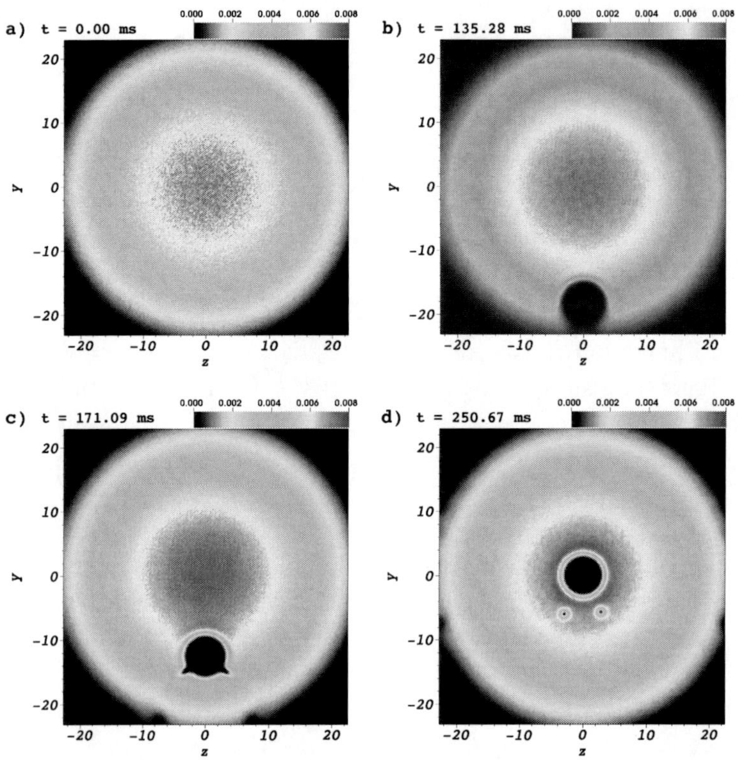

Figure 30. Time evolution of 2D density profile of BEC of dysprosium atoms for $a_s = 66a_0$ and $a_{\rm dd} = 44a_0$, with an obstacle moving at speed $v = v_c = 0.16$ mm/s, when two individual vortices are generated. Each panel shows integrated 2D density profile in y-z plane at different time t. All lengths are expressed in units of μm, and particle density is given in units $N_{\rm at}/l^2$.

tion of the system for velocity $v = v_c = 0.16$ mm/s. The ground state, obtained through imaginary-time propagation is shown in Figure 30(a). The beam, initially outside of the condensate, moves along the y direction, as seen in Figures 30(b) through 30(f). In Figure 30(c) the precursors of the vortices form on the edges of the beam, which then separate from the beam if $v \geq v_c$, a situation seen in Figure 30(d). In Figures 30(e) and 30(f) we see that the generated vortices are stable, and survive for long propagation times.

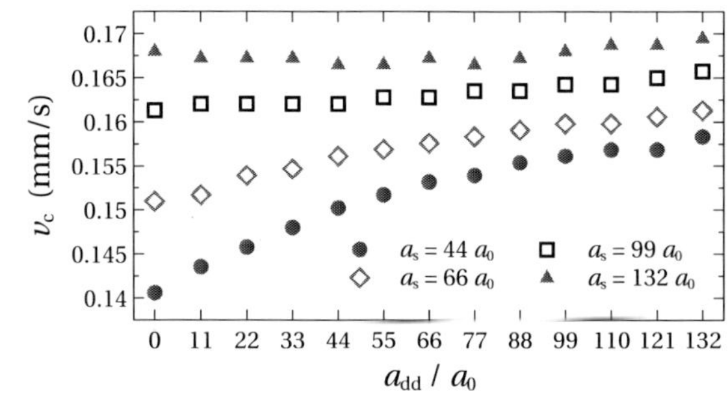

Figure 31. Critical velocity v_c for the emergence of vortices generated by a moving obstacle as a function of the characteristic dipolar interaction length $a_{\rm dd}$ (in units of a_0) for varying values of the s-wave scattering length a_s.

Figure 31 shows results of our numerical study of the dipolar interaction effects on the critical velocity. We see that for large values of the s-wave scattering length, i.e., for the contact interaction comparable or larger than the dipole interaction, effects of decreasing $a_{\rm dd}$ are very small and probably could not be experimentally measured. On the other hand, when contact interaction is tuned down so that the dipole interaction starts to dominate the behavior of the system, critical velocity depends much stronger on $a_{\rm dd}$ and could be easily measured in future experiments.

The above study represents an example on how our programs can be used to verify and compare results of current experiments, as well as to theoretically investigate new phenomena and plan future experiments.

Conclusion

In this chapter we have reviewed an efficient numerical algorithm for solving the nonlinear Schrödinger equation. The stability and efficiency of the used finite difference approach is ensured through properly chosen discretization schemes: the split-step Crank-Nicolson method for the time derivatives and central difference formula for the spatial derivatives. In a computational implementation, the presented numerical algorithm is well suited for various parallelization techniques. We have exploited several parallelization approaches available on present-day computer architectures: OpenMP (for shared memory multiprocessing), OpenMP/MPI (for distributed memory systems, where each compute node uses OpenMP to maximize performance), CUDA (for single GPGPUs), CUDA/MPI (for distributed memory systems, where each compute node has GPGPU installed), as well as their hybrid combinations. For efficient hybrid implementations it is essential to divide the work in a way that is suited to the performances of the available computer architecture. To this purpose additional input parameters are introduced and it is shown that their optimal values can be set using a variant of a genetic algorithm. Both strong and weak scalability of the algorithm are explicitly demonstrated by detailed measurements and subsequent analysis. The scalability is proved to be of high importance as parallel versions of the algorithm allow for efficient studies of realistically large system sizes, which are needed in order to make a direct connection to actual physical experiments. As illustrative examples, three applications of the presented algorithms to the topics of current interest in the field of cold atomic BECs were presented. In section 3 a study of phase separation dynamics of a two–component BEC is performed. In section 4 we addressed excitations of a BEC with the coupling of spin and angular momentum and finally in section 5 we investigated vortex formation dynamics in a stirred BEC in the presence of dipolar interactions.

Acknowledgments

The authors acknowledge valuable discussions with Sadhan K. Adhikari. This work was supported by the Ministry of Education, Science, and Technological Development of the Republic of Serbia under project ON171017. Numerical simulations were performed on the PARADOX supercomputing facility at the Scientific Computing Laboratory of the Institute of Physics Belgrade.

References

[1] Dalfovo, F., Giorgini, S., Pitaevskii, L. P., and Stringari, S. Theory of

Bose-Einstein condensation in trapped gases. *Rev. Mod. Phys. 71* (1999), 463–512.

[2] Gross, E. P. Structure of a quantized vortex in boson systems. *Il Nuovo Cimento 20* (1961), 454–477.

[3] Pitaevskii, L. P. Vortex Lines in an Imperfect Bose Gas. *Sov. Phys. JETP 13* (1961), 451.

[4] Agrawal, G. P. Nonlinear fiber optics: its history and recent progress. *J. Opt. Soc. Am. B 28* (2011), A1.

[5] Taha, T. R., and Ablowitz, M. I. Analytical and Numerical Aspects of Certain Nonlinear Evolution Equations. II. Numerical, Nonlinear Schrödinger Equation. *J. Comput. Phys. 55* (1984), 203–230.

[6] Anderson, M. H., Ensher, J. R., Matthews, M. R., Wieman, C. E., and Cornell, E. A. Observation of Bose-Einstein Condensation in a Dilute Atomic Vapor. *Science 269* (1995), 198–201.

[7] Davis, K. B., Mewes, M.-O., Andrews, M. R., van Druten, N. J., Durfee, D. S., Kurn, D. M., and Ketterle, W. Bose-Einstein condensation in a gas of sodium atoms. *Phys. Rev. Lett. 75* (1995), 3969–3973.

[8] Bradley, C. C., Sackett, C. A., Tollett, J. J., and Hulet, R. G. Evidence of Bose-Einstein Condensation in an Atomic Gas with Attractive Interactions. *Phys. Rev. Lett. 75* (1995), 1687–1690.

[9] Edwards, M., and Burnett, K. Numerical solution of the nonlinear Schrödinger equation for small samples of trapped neutral atoms. *Phys. Rev. A 51* (1995), 1382–1386.

[10] Ruprecht, P. A., Holland, M. J., Burnett, K., and Edwards, M. Time-dependent solution of the nonlinear Schrödinger equation for Bose-condensed trapped neutral atoms. *Phys. Rev. A 51* (1995), 4704–4711.

[11] Dalfovo, F., and Stringari, S. Bosons in anisotropic traps: Ground state and vortices. *Phys. Rev. A 53* (1996), 2477–2485.

[12] Holland, M., and Cooper, J. Expansion of a Bose-Einstein condensate in a harmonic potential. *Phys. Rev. A 53* (1996), R1954–R1957.

[13] Pethick, C. J., and Smith, H. *Bose-Einstein Condensation in Dilute Gases*. Cambridge University Press, 2008.

[14] Pitaevskii, L. P., and Stringari, S. *Bose-Einstein Condensation*. Oxford University Press, 2003.

[15] Snyman, J. *Practical Mathematical Optimization*. Springer, 2005.

[16] Koch, T., Lahaye, T., Metz, J., Fröhlich, B., Griesmaier, A., and Pfau, T. Stabilization of a purely dipolar quantum gas against collapse. *Nat. Phys. 4* (2008), 218–222.

[17] Lu, M., Burdick, N. Q., Youn, S. H., and Lev, B. L. Strongly Dipolar Bose-Einstein Condensate of Dysprosium. *Phys. Rev. Lett. 107* (2011), 190401.

[18] Aikawa, K., Frisch, A., Mark, M., Baier, S., Rietzler, A., Grimm, R., and Ferlaino, F. Bose-Einstein Condensation of Erbium. *Phys. Rev. Lett. 108* (2012), 210401.

[19] Góral, K., and Santos, L. Ground state and elementary excitations of single and binary Bose-Einstein condensates of trapped dipolar gases. *Phys. Rev. A 66* (2002), 023613.

[20] Agrawal, G. P. *Nonlinear Fiber Optics*. Academic Press, 2012.

[21] Antoine, X., Bao, W., and Besse, C. Computational methods for the dynamics of the nonlinear Schrödinger/Gross-Pitaevskii equations. *Comput. Phys. Commun. 184* (2013), 2621–2633.

[22] Muruganandam, P., and Adhikari, S. Fortran programs for the time-dependent Gross-Pitaevskii equation in a fully anisotropic trap. *Comput. Phys. Commun. 180* (2009), 1888–1912.

[23] Vudragović, D., Vidanović, I., Balaž, A., Muruganandam, P., and Adhikari, S. K. C programs for solving the time-dependent Gross-Pitaevskii equation in a fully anisotropic trap. *Comput. Phys. Commun. 183* (2012), 2021–2025.

[24] Press, W. H., Teukolsky, S. A., Vetterling, W. T., and Flannery, B. P. *Numerical Recipes 3rd Edition: The Art of Scientific Computing*, 3 ed. Cambridge University Press, New York, NY, USA, 2007.

[25] Young-S., L. E., Vudragović, D., Muruganandam, P., Adhikari, S. K., and Balaž, A. OpenMP Fortran and C programs for solving the time-dependent Gross-Pitaevskii equation in an anisotropic trap. *Comput. Phys. Commun. 204* (2016), 209–213.

[26] Satarić, B., Slavnić, V., Belić, A., Balaž, A., Muruganandam, P., and Adhikari, S. K. Hybrid OpenMP/MPI programs for solving the time-dependent Gross-Pitaevskii equation in a fully anisotropic trap. *Comput. Phys. Commun. 200* (2016), 411–417.

[27] Frigo, M., and Johnson, S. G. The design and implementation of FFTW3. *Proc. IEEE 93* (2005), 216–231.

[28] Kumar, R. K., Young-S., L. E., Vudragović, D., Balaž, A., Muruganandam, P., and Adhikari, S. Fortran and C programs for the time-dependent dipolar Gross-Pitaevskii equation in an anisotropic trap. *Comput. Phys. Commun. 195* (2015), 117–128.

[29] Lončar, V., Balaž, A., Bogojević, A., Škrbić, S., Muruganandam, P., and Adhikari, S. K. CUDA programs for solving the time-dependent dipolar Gross-Pitaevskii equation in an anisotropic trap. *Comput. Phys. Commun. 200* (2016), 406–410.

[30] Lončar, V., Young-S., L. E., Škrbić, S., Muruganandam, P., Adhikari, S. K., and Balaž, A. OpenMP, OpenMP/MPI, and CUDA/MPI C programs for solving the time-dependent dipolar Gross-Pitaevskii equation. *Comput. Phys. Commun. 209* (2016), 190–196.

[31] Blelloch, G. Scans as primitive parallel operations. *IEEE Trans. Comput. 38* (1989), 1526–1538.

[32] Blelloch, G. E. *Prefix sums and their applications*. Tech. rep., School of Computer Science, Carnegie Mellon University, 1990.

[33] NVIDIA Corporation. *cuFFT Library User's Guide*, 7.5 ed., 2017.

[34] Lee, J., Samadi, M., Park, Y., and Mahlke, S. Transparent cpu-gpu collaboration for data-parallel kernels on heterogeneous systems. In *Proceedings of the 22Nd International Conference on Parallel Architectures and Compilation Techniques* (Piscataway, NJ, USA, 2013), PACT '13, IEEE Press, pp. 245–256.

[35] Papadrakakis, M., Stavroulakis, G., and Karatarakis, A. A new era in scientific computing: Domain decomposition methods in hybrid CPU–GPU architectures. *Comput. Meth. Appl. Mech. Eng. 200* (2011), 1490–1508.

[36] Li, Y., Diamond, J. R., Wang, X., Lin, H., Yang, Y., and Han, Z. Large-scale fast fourier transform on a heterogeneous multi-core system. *Int. J. High Perform. Comput. Appl. 26* (2012), 148–158.

[37] Ogata, Y., Endo, T., Maruyama, N., and Matsuoka, S. An efficient, model-based CPU-GPU heterogeneous FFT library. In *2008 IEEE International Symposium on Parallel and Distributed Processing* (2008), IEEE.

[38] Chen, S., and Li, X. A hybrid GPU/CPU FFT library for large FFT problems. In *2013 IEEE 32nd International Performance Computing and Communications Conference (IPCCC)* (2013), IEEE.

[39] Chu, E., and George, A. *Inside the FFT Black Box.* CRC Press, 1999.

[40] Lončar, V., Young-S., L. E., Škrbić, S., Muruganandam, P., Adhikari, S. K., and Balaž, A. *DBEC-OMP-CUDA-MPI package.* https://data.mendeley.com/datasets/j3z9z379m8/2, 2016.

[41] Lončar, V., Young-S., L. E., Škrbić, S., Muruganandam, P., Adhikari, S. K., and Balaž, A. *Repository of DBEC-GP family of programs.* https://git.ipb.ac.rs/vloncar/DBEC-GP, 2016. Accessed: 2017-04-10.

[42] Lončar, V. Hybrid parallel algorithms for solving nonlinear Schrödinger equation. *PhD Thesis, University of Novi Sad* (2017).

[43] Darken, C., Chang, J., and Moody, J. Learning rate schedules for faster stochastic gradient search. In *Neural Networks for Signal Processing II Proceedings of the 1992 IEEE Workshop* (1992), IEEE.

[44] Schaul, T., Zhang, S., and LeCun, Y. No more pesky learning rates. In *Proceedings of the 30th International Conference on Machine Learning* (Atlanta, Georgia, USA, 2013), S. Dasgupta and D. McAllester, Eds., vol. 28 of *Proceedings of Machine Learning Research*, PMLR, pp. 343–351.

[45] Spall, J. C. Implementation of the simultaneous perturbation algorithm for stochastic optimization. *IEEE Trans. Aerosp. Electron. Syst. 34* (1998), 817–823.

[46] Spall, J. C. Stochastic optimization and the simultaneous perturbation method. In *Proceedings of the 31st conference on Winter simulation Simulation—a bridge to the future - WSC 1999* (1999), ACM Press.

[47] Holland, J. H. *Adaptation in Natural and Artificial Systems.* MIT University Press Group Ltd, 1992.

[48] Mitchell, M. *An Introduction to Genetic Algorithms.* MIT Press, 1998.

[49] Gen, M., and Cheng, R. *Genetic Algorithms and Engineering Optimization.* John Wiley & Sons Inc, 1999.

[50] Kumar, A. Encoding schemes in genetic algorithm. *Int. J. Adv. Res. IT Eng. 2* (2013), 1–7.

[51] Eshelman, L. J., Caruana, R. A., and Schaffer, J. D. Biases in the crossover landscape. In *Proceedings of the Third International Conference on Genetic Algorithms* (San Francisco, CA, USA, 1989), Morgan Kaufmann Publishers Inc., pp. 10–19.

[52] Spears, V. M., and Jong, K. A. D. On the virtues of parameterized uniform crossover. In *Proceedings of the Fourth International Conference on Genetic Algorithms* (1991), pp. 230–236.

[53] Richter, J. N. *On mutation and crossover in the theory of evolutionary algorithms.* PhD thesis, Montana State University, 2010.

[54] White, D. R., and Poulding, S. A rigorous evaluation of crossover and mutation in genetic programming. In *Lecture Notes in Computer Science*, vol. 5481. Springer Berlin Heidelberg, 2009, pp. 220–231.

[55] da Silva, A. P. A., and Falcao, D. M. Fundamentals of genetic algorithms. In *Modern Heuristic Optimization Techniques: Theory and Applications to Power Systems*, K. Y. Lee and M. A. El-Sharkawi, Eds. John Wiley & Sons Inc, 2008, ch. 2, pp. 25–42.

[56] Luke, S., and Spector, L. A comparison of crossover and mutation in genetic programming. In *Genetic Programming 1997: Proceedings of the Second Annual Conference* (1997), J. R. e. a. Koza, Ed., Morgan Kaufmann, pp. 240–248.

[57] Luke, S., and Spector, L. A revised comparison of crossover and mutation in genetic programming. In *Genetic Programming 1998: Proceedings of the Third Annual Conference* (1998), J. R. e. a. Koza, Ed., Morgan Kaufmann, pp. 208–213.

[58] Amdahl, G. M. Validity of the single processor approach to achieving large scale computing capabilities. In *Proceedings of the April 18-20, 1967, Spring Joint Computer Conference - AFIPS '67 (Spring)* (1967), ACM Press.

[59] Open MPI project. *Faq: Running cuda-aware open mpi.* https://www.open-mpi.org/faq/?category=runcuda, 2016. Accessed: 2017-04-10.

[60] Frigo, M., and Johnson, S. G. FFTW *3 documentation*, version 3.3.6. http://www.fftw.org/fftw3_doc/, 2017. Accessed: 2017-04-10.

[61] Edwards, M., Ruprecht, P. A., Burnett, K., Dodd, R. J., and Clark, C. W. Collective Excitations of Atomic Bose-Einstein Condensates. *Phys. Rev. Lett. 77* (1996), 1671–1674.

[62] Vidanović, I., Balaž, A., Al-Jibbouri, H., and Pelster, A. Nonlinear Bose-Einstein-condensate dynamics induced by a harmonic modulation of the s-wave scattering length. *Phys. Rev. A 84* (2011), 013618.

[63] Balaž, A., and Nicolin, A. I. Faraday waves in binary nonmiscible Bose-Einstein condensates. *Phys. Rev. A 85* (2012), 023613.

[64] Balaž, A., Paun, R., Nicolin, A. I., Balasubramanian, S., and Ramaswamy, R. Faraday waves in collisionally inhomogeneous Bose-Einstein condensates. *Phys. Rev. A 89* (2014), 023609.

[65] Sudharsan, J. B., Radha, R., Carina Raportaru, M., Nicolin, A. I., and Balaž, A. Faraday and resonant waves in binary collisionally-inhomogeneous Bose-Einstein condensates. *J. Phys. B-At. Mol. Opt. Phys. 49* (2016), 165303.

[66] Vidanović, I., van Druten, N. J., and Haque, M. Spin modulation instabilities and phase separation dynamics in trapped two-component Bose condensates. *New J. Phys. 15* (2013), 035008.

[67] Al-Jibbouri, H., Vidanović, I., Balaž, A., and Pelster, A. Geometric resonances in Bose-Einstein condensates with two- and three-body interactions. *J. Phys. B-At. Mol. Opt. Phys. 46* (2013), 065303.

[68] Sakhel, R. R., Sakhel, A. R., and Ghassib, H. B. Nonequilibrium Dynamics of a Bose-Einstein Condensate Excited by a Red Laser Inside a Power-Law Trap with Hard Walls. *J. Low Temp. Phys. 173* (2013), 177–206.

[69] Sakhel, R. R., Sakhel, A. R., and Ghassib, H. B. On the phase-correlation and phase-fluctuation dynamics of a strongly excited Bose gas. *Physica B 478* (2015), 68–76.

[70] Sakhel, R. R., Sakhel, A. R., Ghassib, H. B., and Balaz, A. Conditions for order and chaos in the dynamics of a trapped Bose-Einstein condensate in coordinate and energy space. *Eur. Phys. J. D 70* (2016), 66.

[71] Xi, K.-T., Li, J., and Shi, D.-N. Localization of a two-component Bose-Einstein condensate in a one-dimensional random potential. *Physica B 459* (2015), 6–11.

[72] Mithun, T., Porsezian, K., and Dey, B. Disorder-induced vortex lattice melting in a Bose-Einstein condensate. *Phys. Rev. A 93* (2016), 013620.

[73] Gautam, S., and Adhikari, S. K. Fractional-charge vortex in a spinor Bose-Einstein condensate. *Phys. Rev. A 93* (2016), 013630.

[74] Akram, J., and Pelster, A. Sculpting quasi-one-dimensional Bose-Einstein condensate to generate calibrated matter waves. *Phys. Rev. A 93* (2016), 023606.

[75] Akram, J., and Pelster, A. Statics and dynamics of quasi one-dimensional Bose-Einstein condensate in harmonic and dimple trap. *Laser Phys. 26* (2016), 065501.

[76] Sakhel, R. R., and Sakhel, A. R. Application of the Lagrangian variational method to a one-dimensional Bose gas in a dimple trap. *J. Phys. B-At. Mol. Opt. Phys. 50* (2017), 105301.

[77] Akram, J., and Pelster, A. Numerical study of localized impurity in a Bose-Einstein condensate. *Phys. Rev. A 93* (2016), 033610.

[78] Manikandan, K., Muruganandam, P., Senthilvelan, M., and Lakshmanan, M. Manipulating localized matter waves in multicomponent Bose-Einstein condensates. *Phys. Rev E 93* (2016), 032212.

[79] Lai, C.-Y., and Chien, C.-C. Geometry-Induced Memory Effects in Isolated Quantum Systems: Cold-Atom Applications. *Phys. Rev. Applied 5* (2016), 034001.

[80] Akram, J., Girodias, B., and Pelster, A. Quasi one-dimensional Bose-Einstein condensate in a gravito-optical surface trap. *J. Phys. B-At. Mol. Opt. Phys. 49* (2016), 075302.

[81] Khellil, T., Balaž, A., and Pelster, A. Analytical and numerical study of dirty bosons in a quasi-one-dimensional harmonic trap. *New J. Phys. 18* (2016), 063003.

[82] Nakamura, K., Babajanov, D., Matrasulov, D., Kobayashi, M., and Muruganandam, P. Dynamics of trapped interacting vortices in Bose-Einstein condensates: a role of breathing degree of freedom. *J. Phys. A-Math. Theor. 49* (2016), 315102.

[83] Sakhel, R. R., and Sakhel, A. R. Elements of Vortex-Dipole Dynamics in a Nonuniform Bose-Einstein Condensate. *J. Low Temp. Phys. 184* (2016), 1092–1113.

[84] Adhikari, S. K. Statics and dynamics of a self-bound dipolar matter-wave droplet. *Laser Phys. Lett. 14* (2017), 025501.

[85] Adhikari, S. K. Statics and dynamics of a self-bound matter-wave quantum ball. *Phys. Rev. A 95* (2017), 023606.

[86] Adhikari, S. K. Elastic collision and molecule formation of spatiotemporal light bullets in a cubic-quintic nonlinear medium. *Phys. Rev E 94* (2016), 032217.

[87] Gautam, S., and Adhikari, S. K. Phase separation in a spin-orbit-coupled Bose-Einstein condensate. *Phys. Rev. A 90* (2014), 043619.

[88] Gautam, S., and Adhikari, S. K. Mobile vector soliton in a spin-orbit coupled spin-1 condensate. *Laser Phys. Lett. 12* (2015), 045501.

[89] Gautam, S., and Adhikari, S. K. Vector solitons in a spin-orbit-coupled spin-2 Bose-Einstein condensate. *Phys. Rev. A 91* (2015), 063617.

[90] Gautam, S., and Adhikari, S. K. Vortex-bright solitons in a spin-orbit-coupled spin-1 condensate. *Phys. Rev. A 95* (2017), 013608.

[91] Vinayagam, P. S., Radha, R., Bhuvaneswari, S., Ravisankar, R., and Muruganandam, P. Bright soliton dynamics in spin orbit-Rabi coupled Bose-Einstein condensates. *Commun. Nonlinear Sci. Numer. Simulat. 50* (2017), 68–76.

[92] Vasić, I., and Balaž, A. Excitation spectra of a Bose-Einstein condensate with an angular spin-orbit coupling. *Phys. Rev. A 94* (2016), 033627.

[93] Kishor Kumar, R., and Muruganandam, P. Vortex dynamics of rotating dipolar Bose-Einstein condensates. *J. Phys. B-At. Mol. Opt. Phys. 45* (2012), 215301.

[94] Kishor Kumar, R., and Muruganandam, P. Numerical studies on vortices in rotating dipolar Bose-Einstein condensates. In *J. Phys.: Conf. Ser.* (2014), vol. 497, p. 012036.

[95] Kumar, R. K., and Muruganandam, P. Effect of optical lattice potentials on the vortices in rotating dipolar Bose-Einstein condensates. *Eur. Phys. J. D 68* (2014), 289.

[96] Kishor Kumar, R., Sriraman, T., Fabrelli, H., Muruganandam, P., and Gammal, A. Three-dimensional vortex structures in a rotating dipolar Bose-Einstein condensate. *J. Phys. B-At. Mol. Opt. Phys. 49* (2016), 155301.

[97] Young-S., L. E., and Adhikari, S. K. Mixing, demixing, and structure formation in a binary dipolar Bose-Einstein condensate. *Phys. Rev. A 86* (2012), 063611.

[98] Young-S., L. E., and Adhikari, S. K. Dipolar droplet bound in a trapped Bose-Einstein condensate. *Phys. Rev. A 87* (2013), 013618.

[99] Adhikari, S. K. Stability of trapped degenerate dipolar Bose and Fermi gases. *J. Phys. B-At. Mol. Opt. Phys. 46* (2013), 115301.

[100] Adhikari, S. K. Stability and collapse of fermions in a binary dipolar boson-fermion ^{164}Dy-^{161}Dy mixture. *Phys. Rev. A 88* (2013), 043603.

[101] Kishor Kumar, R., Muruganandam, P., and Malomed, B. A. Vortical and fundamental solitons in dipolar Bose-Einstein condensates trapped in isotropic and anisotropic nonlinear potentials. *J. Phys. B-At. Mol. Opt. Phys. 46* (2013), 175302.

[102] Adhikari, S. K. Stable and mobile excited two-dimensional dipolar Bose-Einstein condensate solitons. *J. Phys. B-At. Mol. Opt. Phys. 47* (2014), 225304.

[103] Adhikari, S. K. Stable, mobile, dark-in-bright, dipolar Bose-Einstein-condensate solitons. *Phys. Rev. A 89* (2014), 043615.

[104] Adhikari, S. K. Demixing and symmetry breaking in binary dipolar Bose-Einstein-condensate solitons. *Phys. Rev. A 89* (2014), 013630.

[105] Adhikari, S. K. Bright dipolar Bose-Einstein-condensate soliton mobile in a direction perpendicular to polarization. *Phys. Rev. A 90* (2014), 055601.

[106] Adhikari, S. K. Two-dimensional bright and dark-in-bright dipolar Bose-Einstein condensate solitons on a one-dimensional optical lattice. *Laser Phys. Lett. 13* (2016), 085501.

[107] Adhikari, S. K. Stable and mobile two-dimensional dipolar ring-dark-in-bright Bose-Einstein condensate soliton. *Laser Phys. Lett. 13* (2016), 035502.

[108] Adhikari, S. K. Self-trapping of a dipolar Bose-Einstein condensate in a double well. *Phys. Rev. A 89* (2014), 043609.

[109] Ho, T.-L., and Shenoy, V. B. Binary Mixtures of Bose Condensates of Alkali Atoms. *Phys. Rev. Lett. 77* (1996), 3276–3279.

[110] Matthews, M. R., Hall, D. S., Jin, D. S., Ensher, J. R., Wieman, C. E., Cornell, E. A., Dalfovo, F., Minniti, C., and Stringari, S. Dynamical

Response of a Bose-Einstein Condensate to a Discontinuous Change in Internal State. *Phys. Rev. Lett. 81* (1998), 243–247.

[111] Hall, D. S., Matthews, M. R., Ensher, J. R., Wieman, C. E., and Cornell, E. A. Dynamics of Component Separation in a Binary Mixture of Bose-Einstein Condensates. *Phys. Rev. Lett. 81* (1998), 1539–1542.

[112] Miesner, H.-J., Stamper-Kurn, D. M., Stenger, J., Inouye, S., Chikkatur, A. P., and Ketterle, W. Observation of Metastable States in Spinor Bose-Einstein Condensates. *Phys. Rev. Lett. 82* (1999), 2228–2231.

[113] Salasnich, L., Parola, A., and Reatto, L. Effective wave equations for the dynamics of cigar-shaped and disk-shaped Bose condensates. *Phys. Rev. A 65* (2002), 043614.

[114] Pu, H., and Bigelow, N. P. Collective Excitations, Metastability, and Nonlinear Response of a Trapped Two-Species Bose-Einstein Condensate. *Phys. Rev. Lett. 80* (1998), 1134–1137.

[115] Timmermans, E. Phase Separation of Bose-Einstein Condensates. *Phys. Rev. Lett. 81* (1998), 5718–5721.

[116] Kasamatsu, K., and Tsubota, M. Multiple Domain Formation Induced by Modulation Instability in Two-Component Bose-Einstein Condensates. *Phys. Rev. Lett. 93* (2004), 100402.

[117] Kasamatsu, K., and Tsubota, M. Modulation instability and solitary-wave formation in two-component Bose-Einstein condensates. *Phys. Rev. A 74* (2006), 013617.

[118] Navarro, R., Carretero-González, R., and Kevrekidis, P. G. Phase separation and dynamics of two-component Bose-Einstein condensates. *Phys. Rev. A 80* (2009), 023613.

[119] Lin, Y.-J., Jiménez-García, K., and Spielman, I. B. Spin-orbit-coupled Bose-Einstein condensates. *Nature (London) 471* (2011), 83–86.

[120] Beeler, M. C., Williams, R. A., Jiménez-García, K., Leblanc, L. J., Perry, A. R., and Spielman, I. B. The spin Hall effect in a quantum gas. *Nature (London) 498* (2013), 201–204.

[121] Zhang, J.-Y., Ji, S.-C., Chen, Z., Zhang, L., Du, Z.-D., Yan, B., Pan, G.-S., Zhao, B., Deng, Y.-J., Zhai, H., Chen, S., and Pan, J.-W. Collective Dipole Oscillations of a Spin-Orbit Coupled Bose-Einstein Condensate. *Phys. Rev. Lett. 109* (2012), 115301.

[122] Khamehchi, M. A., Zhang, Y., Hamner, C., Busch, T., and Engels, P. Measurement of collective excitations in a spin-orbit-coupled Bose-Einstein condensate. *Phys. Rev. A 90* (2014), 063624.

[123] Cheuk, L. W., Sommer, A. T., Hadzibabic, Z., Yefsah, T., Bakr, W. S., and Zwierlein, M. W. Spin-injection spectroscopy of a spin-orbit coupled fermi gas. *Phys. Rev. Lett. 109* (2012), 095302.

[124] Li, J., Huang, W., Shteynas, B., Burchesky, S., Top, F. C., Su, E., Lee, J., Jamison, A. O., and Ketterle, W. Spin-Orbit Coupling and Spin Textures in Optical Superlattices. *Phys. Rev. Lett. 117* (2016), 185301.

[125] Li, J.-R., Lee, J., Huang, W., Burchesky, S., Shteynas, B., Top, F. Ç., Jamison, A. O., and Ketterle, W. A stripe phase with supersolid properties in spin-orbit-coupled Bose-Einstein condensates. *Nature (London) 543* (2017), 91–94.

[126] Galitski, V., and Spielman, I. B. Spin-orbit coupling in quantum gases. *Nature (London) 494* (2013), 49–54.

[127] Zhai, H. Degenerate quantum gases with spin-orbit coupling: a review. *Rep. Prog. Phys. 78* (2015), 026001.

[128] Hu, Y.-X., Miniatura, C., and Grémaud, B. Half-skyrmion and vortex-antivortex pairs in spinor condensates. *Phys. Rev. A 92* (2015), 033615.

[129] DeMarco, M., and Pu, H. Angular spin-orbit coupling in cold atoms. *Phys. Rev. A 91* (2015), 033630.

[130] Qu, C., Sun, K., and Zhang, C. Quantum phases of Bose-Einstein condensates with synthetic spin-orbital-angular-momentum coupling. *Phys. Rev. A 91* (2015), 053630.

[131] Sun, K., Qu, C., and Zhang, C. Spin-orbital-angular-momentum coupling in Bose-Einstein condensates. *Phys. Rev. A 91* (2015), 063627.

[132] Pitaevskii, L. P., and Rosch, A. Breathing modes and hidden symmetry of trapped atoms in two dimensions. *Phys. Rev. A 55* (1997), R853–R856.

[133] Kwon, W. J., Moon, G., Seo, S. W., and Shin, Y. Critical velocity for vortex shedding in a Bose-Einstein condensate. *Phys. Rev. A 91* (2015), 053615.

[134] Kadau, H., Schmitt, M., Wenzel, M., Wink, C., Maier, T., Ferrier-Barbut, I., and Pfau, T. Observing the Rosensweig instability of a quantum ferrofluid. *Nature (London) 530* (2016), 194–197.

INDEX

A

algorithm, viii, ix, 63, 64, 65, 71, 72, 73, 74, 75, 76, 80, 81, 82, 83, 84, 85, 86, 87, 88, 89, 90, 91, 92, 93, 97, 99, 102, 103, 104, 105, 106, 110, 113, 117, 119, 122, 143, 144, 148, 149
Amdahl's law, 99
amplitude, 124, 132
annealing, 88
atoms, ix, 64, 65, 66, 67, 68, 121, 122, 123, 126, 134, 135, 136, 137, 138, 139, 140, 141, 142, 145, 156

B

bipartite network, viii, 33, 35, 38, 41, 42, 44, 45, 46, 47, 55, 59
Bose-Einstein condensate (BEC), viii, ix, 63, 64, 66, 67, 68, 121, 122, 123, 125, 126, 129, 133, 134, 135, 136, 137, 138, 139, 140, 141, 142, 144, 145, 146, 150, 151, 152, 153, 154, 155, 156, 157
boson, 144, 152, 154
breathing mode, 121, 123, 129, 130, 131
brute-force search, 85

C

classes, 6, 8, 9, 14, 15, 17, 18, 19, 20, 30
clustering, 43, 44, 47, 48
clusters, 73, 82, 110
code generation, 9
cold atoms, 65, 66, 67, 68, 121, 156
collective dynamics, viii, 33, 48, 49, 55
collective modes, 121
communication, 34, 94, 108, 112, 113, 114, 119
communication overhead, 113
communication technologies, 34
complex network theory, 34, 35, 55
complexity, 2, 8, 71, 76, 80
computation, 2, 5, 14, 18, 25, 28, 30, 76, 77, 78, 80, 81, 84, 88, 102, 105, 107, 108, 117
computational modeling, viii, 33
computational social science, 34, 56
computing, vii, ix, 1, 2, 7, 20, 21, 29, 64, 71, 77, 83, 84, 86, 92, 93, 94, 95, 107, 110, 150
configuration, 39, 48, 107, 114, 122, 127, 128
connectivity patterns, 42
CPU, 73, 76, 77, 78, 80, 81, 82, 83, 84, 85, 86, 93, 94, 98, 102, 107, 110, 120, 147, 148
crossover operator, 90, 91
CUDA, ix, 29, 64, 72, 73, 76, 82, 94, 95, 96, 102, 103, 104, 105, 106, 107, 108, 109, 110, 112, 116, 117, 119, 120, 144, 147, 148

D

data distribution, 77, 80, 83, 84
data structure, 89
data transfer, 76, 84, 85, 112, 117

density profile, 120, 136, 137, 138, 139, 141, 142
dipole mode, 129, 131
dipole-dipole interaction, ix, 64, 140
dipoles, 68, 133, 134, 135, 138, 140
discrete Fourier transform (DFT), 74, 75, 80, 81, 82
discretization scheme, 69, 73, 143
discretization, viii, 64, 69, 70, 71, 73, 74, 143
distributed data transpose, 80
distributed memory, ix, 64, 71, 73, 74, 77, 82, 83, 117, 143, 144
dysprosium, 134, 140, 141, 142

E

elitist selection, 90
encoding scheme, 149
energy, 66, 67, 68, 71, 110, 122, 127, 128, 151
energy conservation, 68
energy efficiency, 110
event-driven dynamics, 45, 46, 47, 49
evolution, 34, 35, 43, 44, 45, 46, 49, 53, 54, 57, 60, 61, 65, 66, 67, 89, 91, 124, 130, 136, 137, 138, 139, 141, 142
evolutionary computation, 83

F

Fast Fourier transform (FFT), 25, 26, 27, 74, 80, 81, 82, 85, 92, 97, 103, 105, 114, 116, 117, 120, 148
FFTW, 72, 76, 80, 83, 94, 103, 105, 113, 114, 116, 119, 120, 150
fiber optics, 68, 145
formation, ix, 56, 64, 122, 133, 134, 135, 140, 141, 144, 152, 153, 155
formula, 71, 143
fractal structure, 51
function values, 74, 75, 78
functional programming, 4, 16, 18

G

Gaussian laser beam, 133, 135, 141
genetic algorithm, 85, 88, 144, 149
genetic programming, 149, 150
geometry, 121, 122, 123
Google, 9, 26, 29
gradient descent method, 67
Groovy, vii, 1, 2, 3, 4, 5, 7, 8, 9, 11, 12, 13, 14, 15, 16, 17, 18, 19, 20, 21, 22, 23, 24, 25, 26, 27, 28, 29, 30
GroovyLab, vii, 1, 2, 3, 4, 5, 7, 8, 9, 11, 12, 13, 14, 15, 18, 19, 20, 21, 23, 24, 25, 26, 27, 28, 29
Gross-Pitaevskii equation, 63, 64, 129, 146, 147
group activities, 47, 52, 53
group size, 36, 57
grouping, 85

H

half-skyrmion, 129
Hamiltonian, 74, 75, 78, 79, 122, 125
heterogeneity, 41, 46, 47, 55
heterogeneous systems, 147
host, 78, 79, 80, 81, 82, 90, 104, 106, 107, 116, 117, 120
hybrid, ix, 64, 71, 72, 73, 74, 76, 77, 78, 80, 82, 83, 84, 85, 88, 92, 94, 95, 104, 105, 106, 107, 108, 109, 110, 113, 114, 117, 120, 144, 147, 148
hybrid FFT, 105, 114

I

imaginary-time propagation, 74, 75, 102, 106, 107, 110, 135, 142
initial state, 123, 128, 137, 141
integration, viii, ix, 64, 69, 127
interface, 3, 4, 6, 23, 28, 29, 76
invitation to participate, 36
iteration, 74, 75, 85, 86, 87, 88, 93, 94, 95, 97, 99, 100, 102, 103, 104, 105, 108, 109, 114, 117, 136

J

Java, vii, 1, 2, 3, 4, 5, 6, 7, 8, 9, 11, 12, 13, 15, 16, 17, 18, 20, 21, 22, 23, 24, 25, 26, 27, 28, 29, 30

L

languages, vii, 1, 3, 5, 14, 15, 16, 20, 29, 31
learning rate, 88, 92, 148

M

mapping, 11, 38
matrix, 4, 5, 6, 7, 8, 9, 10, 12, 13, 25, 27, 28, 29, 38, 41, 81, 96, 122, 125
Meetup, vii, viii, 33, 35, 36, 37, 38, 39, 40, 41, 42, 43, 44, 45, 46, 47, 48, 49, 50, 51, 53, 54, 55, 56, 57, 59, 61
memory, ix, 51, 52, 55, 60, 64, 71, 72, 73, 76, 77, 78, 79, 80, 81, 82, 94, 102, 103, 104, 107, 116, 117, 120, 143
message passing, 113
methodology, 56, 83, 97, 107, 134, 140
Microsoft, 28
Ministry of Education, 56, 144
mixing, 42, 43, 45, 47
models, 14, 57, 60, 86, 103, 119
modifications, 23, 76
modulation instability, 124
MPI, ix, 64, 71, 72, 73, 82, 83, 85, 93, 94, 95, 96, 107, 108, 109, 110, 111, 112, 113, 114, 115, 116, 117, 118, 119, 120, 133, 143, 144, 147, 148, 150
mutation, 89, 90, 91, 92, 149, 150
mutation operator, 90, 91
mutation rate, 92
MySpace, 57

N

network density, 41
network theory, 34, 35, 55, 58
nodes, ix, 20, 38, 42, 43, 44, 45, 46, 47, 48, 64, 73, 77, 83, 93, 94, 107, 108, 110, 111, 112, 114, 116, 117, 119
nonlinear optics, 65
nonlinear Schrödinger equation, viii
numerical analysis, 6, 29

O

OpenMP, ix, 64, 71, 73, 76, 82, 94, 95, 96, 97, 98, 99, 100, 101, 102, 104, 105, 106, 107, 108, 109, 110, 112, 113, 114, 116, 117, 119, 120, 143, 144, 146, 147
operations, 6, 7, 9, 13, 14, 22, 23, 26, 29, 84, 94, 100, 102, 147
optical lattice, 121, 153, 154
optimization, ix, 64, 71, 85, 86, 88, 89, 92, 93, 105, 106, 107, 112, 113, 117, 120, 148, 149
optimization method, 88, 89, 92, 93, 105, 112, 120
orbit, 121, 125, 130, 152, 153, 155, 156
oscillation, 123, 124, 131, 132
overlap, 54, 84, 116, 117, 122, 123, 128

P

parallel algorithm, 148
parallel implementation, 73, 83
parallelization, ix, 64, 65, 71, 73, 75, 76, 77, 143
parameter set, 85, 92
partition, 38, 39, 42, 44, 45, 46, 47, 55
physical phenomena, viii, 63, 65, 140
physical sciences, viii, 63
physics, 34, 35, 55, 56, 60, 65, 122
polarization, 68, 154
probability, 36, 37, 39, 40, 41, 42, 43, 45, 52, 55, 90, 130
probability distribution, 41, 45, 130
professional development, 37
programming, vii, viii, ix, 1, 2, 3, 16, 17, 20, 30, 64, 71, 72, 103
programming languages, 16, 72
propagation, viii, 63, 67, 68, 69, 74, 75, 76, 78, 79, 80, 85, 86, 87, 92, 93, 94, 97, 100, 102, 103, 104, 105, 106, 107, 109, 110, 120, 126, 127, 135, 137, 141, 142

Q

quantitative technique, 34
quantum mechanics, 65, 66

R

random walk, 90
real time, 50, 53, 55, 94
recombination, 90
reproduction, 89, 90
requirement, 100, 108
resources, 34, 83, 95, 108
roulette-wheel selection, 89
row-column algorithm, 8, 10

S

Scala, vii, 1, 2, 3, 4, 5, 6, 7, 8, 9, 10, 11, 12, 13, 14, 15, 16, 17, 18, 19, 20, 21, 22, 23, 24, 25, 26, 28, 29, 30
ScalaLab, vii, 1, 2, 3, 4, 5, 7, 8, 9, 10, 12, 13, 14, 18, 20, 21, 23, 24, 25, 26, 27, 28, 29, 30
scaling, 53, 83, 94, 95, 97, 98, 99, 100, 101, 102, 108, 109, 111, 112, 113, 114, 115, 116, 117, 118, 119
scaling efficiency, 94, 95, 97, 98, 100, 101, 102, 111, 116, 117, 119
scaling law, 53
scattering, 134, 140, 143, 150
Schrödinger equation, viii
scientific computing, vii, 29, 147
scripts, 2, 8, 19, 25, 28, 29
search space, 85, 89, 90
segregation, 34, 57
selection operator, 89
semi-implicit Crank-Nicolson method, 64, 72, 143
slab decomposition, 77
social network, viii, 33, 35, 43, 46, 47, 48, 55, 56, 58
software, 2, 26, 29, 60, 87
solitons, 121, 153, 154
solution, 28, 67, 68, 70, 71, 75, 76, 86, 88, 89, 90, 91, 92, 93, 128, 133, 145
speedup, 94, 95, 97, 98, 99, 102, 103, 105, 106, 107, 108, 109, 110, 111, 112, 113, 119
spin-orbit coupling, 153, 156
split-step approximation, 69
standard deviation, 45, 46, 47, 51, 53
structure, vii, viii, 18, 33, 34, 35, 36, 42, 43, 44, 46, 48, 49, 53, 55, 56, 57, 60, 153
subgroups, 48, 54, 55
subnetworks, 49

T

techniques, vii, viii, ix, 8, 33, 34, 35, 36, 64, 65, 71, 83, 90, 91, 143
technology, 2, 30, 37
testing, 25, 83, 92, 96, 100, 108, 120
time series, 35, 49, 50, 51, 53, 54
tournament selection, 89, 90, 92
transformations, 20, 21, 23, 72, 126

V

variations, 25, 27
vector, 11, 12, 65, 68, 74, 83, 86, 152
velocity, 134, 135, 138, 139, 140, 141, 143, 157
vortex, 121, 122, 127, 128, 129, 131, 132, 133, 134, 135, 138, 139, 140, 141, 144, 151, 153, 157

W

workload, 100, 101, 114, 120
World Wide Web (WWW), 58
WWW, 58